A History of Science in Society

A History of Science in Society

Volume I:
From the Ancient Greeks
to the Scientific Revolution

Third Edition

Andrew Ede and Lesley B. Cormack

UNIVERSITY OF TORONTO PRESS

utppublishing.com

Library and Archives Canada Cataloguing in Publication

Ede, Andrew

[History of science in society (Toronto, Ont.)]

A history of science in society / Andrew Ede and Lesley B. Cormack.

Kept up to date by replacement editions.

Includes bibliographical references and index.

Issued in print and electronic formats.

ISBN 978-1-4426-3503-6 (v. 1, third edition : paperback).—ISBN 978-1-4426-3504-3 (v. 1, third edition : epub). —ISBN 978-1-4426-3505-0 (v. 1, third edition : pdf).

1. Science—History. 2. Science—Social aspects—History. I. Cormack, Lesley B., 1957–, author II. Title.

Q125.E33 2012b   509   C2012–900477–4

C2016–905501–9

We welcome comments and suggestions regarding any aspect of our publications—please feel free to contact us at news@utphighereducation.com or visit our Internet site at www.utppublishing.com.

North America

5201 Dufferin Street

North York, Ontario, Canada, M3H 5T8

2250 Military Road

Tonawanda, New York, USA, 14150

ORDERS PHONE: 1-800-565-9523

ORDERS FAX: 1-800-221-9985

ORDERS E-MAIL: utpbooks@utpress.utoronto.ca

UK, Ireland, and continental Europe

5201 Dufferin Street

Estover Road, Plymouth, PL6 7PY, UK

ORDERS PHONE: 44 (0) 1752 202301

ORDERS FAX: 44 (0) 1752 202333

ORDERS E-MAIL: enquiries@nbninternational.com

Every effort has been made to contact copyright holders; in the event of an error or omission, please notify the publisher.

The University of Toronto Press acknowledges the financial support for its publishing activities of the Government of Canada through the Canada Book Fund.

Printed in the United States of America.

CONTENTS

ILLUSTRATIONS

Figures

Plates

CONNECTIONS
BOXES

ACKNOWLEDGMENTS

To Graham and Quin, who have grown up with this book—and who put up with two authors working in the house at the same time. We would also like to thank those people who helped make this book possible: our editor and publisher; friends and colleagues who read early drafts and gave advice; reviewers and users who have offered helpful criticism and forced us to defend our position; and all the amazing historians of science on whose shoulders (or toes) we stand.

INTRODUCTION

S cience has transformed human history. It has changed how we see the universe, how we interact with nature and each other, and how we live our lives. It may, in the future, even change what it means to be human. The history of such a powerful force deserves a full and multifaceted examination. Yet a history of science is unlike a history of monarchs, generals, steam engines, or wars because science isn't a person, an object, or an event. It is an idea, the idea that humans can understand the physical world.

This is a history of what happens when a legion of thinkers, at different times and from different backgrounds, turned their minds and hands to the investigation of nature. In the process, they transformed the world.

The history of science is such a vast subject that no single book about it can really be comprehensive, and so the story we tell examines science from a particular point of view. Some histories of science have focused on the intellectual development of ideas, while others have traced the course of particular subjects such as astronomy or physics. In this book, we have chosen to look at science from two related perspectives that we believe offer a window onto the historical processes that shaped the study of nature. First, we have examined the link between the philosophical pursuit of knowledge and the desire of both the researchers and their supporters to make that knowledge useful. There has always been a tension between the intellectual aspects of science and the application of scientific knowledge. The ancient Greek philosophers struggled with this problem, and it is still being debated today. The call in every age by philosophers and scientists for more support for "research for its own sake" is indicative of the tension between the search for knowledge and the pressure to apply that knowledge. What counts as useful knowledge differed from patron to patron and society to society, so that the Grand Duke Cosimo de' Medici and the United States Department of Energy looked for quite different "products" to be created by their clients, but both traded support for the potential of utility.

The tension between intellectual pursuits and demands for some kind of product not only was felt by many natural philosophers and scientists but has also led to controversy among historians of science. Where does science end and technology begin? they have asked. Perhaps the most famous articulation of this is the "scholar and craftsman debate." Historians of science have tried to understand the relationship between those people primarily interested in the utility of knowledge (the craftsmen) and those interested in the intellectual understanding of the world (the scholars). Some historians have denied the connection, but we feel it is integral to the pursuit of natural knowledge. The geographers of the early modern period provide a good example of the necessity of this interconnection. They brought the skills of the navigator together with the abstract knowledge of the mathematician. Translating the spherical Earth onto flat maps was an intellectual challenge, while tramping to the four corners of the globe to take measurements was an extreme physical challenge. Getting theory and practice right could mean the difference between profit or loss, or even life and death.

Our second aim has been to trace the history of science by its social place. Science does not exist in disembodied minds, but is part of living, breathing society. It is embedded in institutions such as schools, princely courts, government departments, and even in the training of soldiers. As such, we have tried to relate scientific work to the society in which it took place, tracing the interplay of social interest with personal interest. This has guided our areas of emphasis so that, for example, we give alchemy a greater allocation of space than some other histories of science because it was more socially significant than topics such as astronomy or physics in the same period. There were far more alchemists than astronomers, and they came from all ranks and classes of people, from peasants to popes. In the longer term, the transformation of alchemy into chemistry had a very great impact on the quality of everyday life. This is not to say that we neglect astronomy or physics, but rather that we have tried to focus on what was important to the people of the era and to avoid projecting the importance of later work on earlier ages.

In each chapter, we have highlighted one aspect of this interaction of science and society, from politics and religion to economics and warfare, under the heading

"Connections." While each of these vignettes is part of the larger narrative of the book, they can also be read as individual case studies.

It is from the two perspectives of utility and social place that our subtitle comes. As we began to look at the work of natural philosophers and scientists over more than 2,000 years, we found ourselves more and more struck by the consistency of the issue of the utility of knowledge. Plato disdained the utility of knowledge, but he promoted an understanding of geometry. Eratosthenes used geometry to measure the diameter of the Earth, which had many practical applications. In the modern era, we have seen many cases of scientific work unexpectedly turned into consumer goods. The cathode ray tube, for instance, was a device created to study the nature of matter, but it ended up in the heart of the modern television. Philosophers and scientists have always walked a fine line between the role of intellectual and the role of technician. Too far to the technical side and a person will appear to be an artisan and lose their status as an intellectual. Too far to the intellectual side, a person will have trouble finding support because they have little to offer potential patrons.

Although the tension over philosophy and utility has always existed for the community of researchers, we did not subtitle our book "Philosophy *and* Utility." This is because the internal tension was not the only aspect of philosophy and utility that we saw over time. Natural philosophy started as an esoteric subject studied by a small, often very elite, group of people. Their work was intellectually important but had limited impact on the wider society. Over time, the number of people interested in natural philosophy grew, and as the community grew, so too did the claims of researchers that what they were doing would benefit society. Through the early modern and modern eras, scientists increasingly promoted their work on the basis of its potential utility, whether as a cure for cancer or as a better way to cook food. And, in large part, the utility of science has been graphically demonstrated in everything from the production of colour-fast dyes to the destruction of whole cities with a single bomb. We have come to expect science to produce things we can use, and, further, we need scientifically trained people to keep our complex systems working—everything from testing the purity of our drinking water to teaching science in school. Our subtitle reflects the changing social expectation of science.

We have also made some choices about material based on the need for brevity. This book could not include all historical aspects of all topics in science or even introduce all the disciplines in science. We picked examples that illustrate key events and ideas rather than give exhaustive detail. For instance, the limited amount of medical history we include looks primarily at examples from medicine that treated the body as an object of research and thus as part of a larger intellectual movement in natural philosophy. We also chose to focus primarily on Western developments in natural philosophy and science, although we tried to acknowledge that natural philosophy existed in other places as well and that Western science did not develop in isolation. Especially in the early periods, Western thinkers were absorbing ideas, materials, and information from a wide variety of sources. By the seventeenth and eighteenth centuries, Western scholars were interacting with other cultures and exchanging information, although not on an equal footing. In later periods, Western science became a powerful tool for modernization and internationalization of countries around the world. *A History of Science* tells a particular—and important—story about the development of this powerful part of human culture, which has and continues to transform all our lives. To study the history of science is to study one of the great threads in the cloth of human history.

**CHAPTER
TIMELINE**

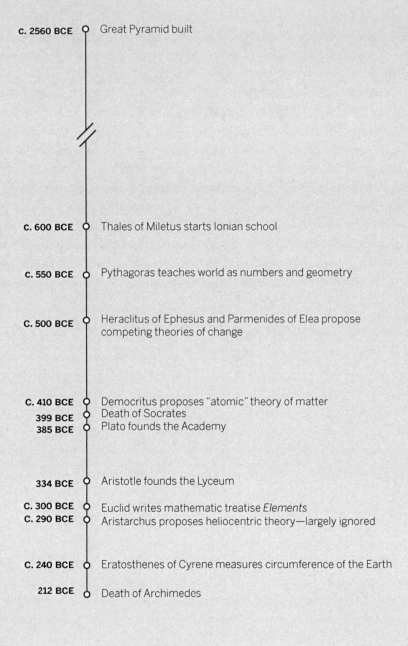

c. 2560 BCE — Great Pyramid built

c. 600 BCE — Thales of Miletus starts Ionian school

c. 550 BCE — Pythagoras teaches world as numbers and geometry

c. 500 BCE — Heraclitus of Ephesus and Parmenides of Elea propose competing theories of change

c. 410 BCE — Democritus proposes "atomic" theory of matter
399 BCE — Death of Socrates
385 BCE — Plato founds the Academy

334 BCE — Aristotle founds the Lyceum

c. 300 BCE — Euclid writes mathematic treatise *Elements*
c. 290 BCE — Aristarchus proposes heliocentric theory—largely ignored

c. 240 BCE — Eratosthenes of Cyrene measures circumference of the Earth

212 BCE — Death of Archimedes

THE ORIGINS OF
NATURAL PHILOSOPHY

1

The roots of modern science are found in the heritage of natural philosophy created by a small group of ancient Greek philosophers. The voyage from the Greeks to the modern world was a convoluted one, and natural philosophy was transformed by the cultures that explored and re-explored the foundational ideas of those Greek thinkers. Despite intellectual and practical challenges, the Greek conceptions of how to think about the world and how the universe worked remained at the heart of any investigation of nature in Europe and the Middle East for almost 2,000 years. Even when natural philosophers began to reject the conclusions of the Greek philosophers, the rejection itself still carried with it the form and concerns of Greek philosophy. Today, when virtually nothing of Greek method or conclusions about the physical world remains, the philosophical concerns about how to understand what we think we know about the universe still echo in our modern version of natural philosophy.

To understand why Greek natural philosophy was such an astounding achievement, we must consider the conditions that led to the creation of a philosophy of nature. Since the earliest times of human activity, the observation of nature has been a key to human survival. Knowledge of everything—from which plants are edible to where babies come from—was part of the knowledge acquired and passed down through the generations. In addition to practical knowledge useful for daily life, humans worked to understand the nature of existence and encapsulated their knowledge and conclusions in a framework of mytho-poetic stories.

Humans have always wanted to know more than just what is in the world; they want to know why the world is the way it is.

Early Civilizations and the Development of Knowledge

With the rise of agriculture and the development of urban civilization, the types of knowledge about nature were diversified as new skills were created. There arose four great cradles of civilization along the river systems of the Nile, the Tigris-Euphrates, the Indus-Ganges, and the Yellow. They shared the common characteristic of a large river that was navigable over a long distance and that flooded the region on a periodic basis. The Nile in particular flooded so regularly that its rise and fall was part of the timekeeping of the Egyptians. These floods renewed the soil, and the lands in temperate to subtropical zones were (and are) agriculturally abundant, providing food to support large populations.

A growing group of people were freed from farm work by the surplus the land provided. These people were the artisans, soldiers, priests, nobles, and bureaucrats who could turn their efforts to the development and running of an empire. The mastery of these skills required increasingly longer periods of study and practice. Artisans required apprenticeships to acquire and master their arts, while the priest class took years to learn the doctrine and methods of correct observance. The military and ruling classes required training from childhood to grow proficient in their duties. Because the empires were long-lasting, especially the Egyptian empire, the rulers planned for the long term, thinking not just about the present season but about the years ahead and even generations into the future. Thus, these civilizations could take on major building projects such as the Great Wall of China or the Great Pyramid of Giza.

In addition to the obvious agricultural and economic advantage provided by the rivers, they had a number of subtle effects on the intellectual development of ancient civilizations. Dealing with large-scale agricultural production required counting and measurement of length, weight, area, and volume, and that led to accounting skills and record-keeping. Agriculture and religion were intertwined, and both depended on timekeeping to organize activities necessary for worship and production, which in turn led to astronomical observation and calendars. As these societies moved from villages to regional kingdoms and finally became empires, record-keeping exceeded what could be left to memory. Writing and accounting developed to deal with the problems of remembering and recording the myriad activities of complex religions, government bureaucracies, and the decisions of judges at courts of law.

Another aspect of intellectual development that came from the periodic flooding had to do with the loss of local landmarks, so skills of surveying were developed. Rather than setting the boundaries of land by objects such as trees or rocks, which changed with every inundation, the land was measured from objects unaffected by the flooding. In addition to the practical skills of land measurement, surveying also introduced concepts of geometry and the use of level and angle measuring devices. These were then used for building projects such as irrigation systems, canals, and large buildings. In turn, surveying tools were closely related to the tools used for navigation and astronomy.

These kinds of practical skills contributed to a conception of the world based on abstract models. In other words, counting cattle contributed to the concept of arithmetic as a subject that could be taught independent of any actual object to be counted. Similarly, getting from place to place by boat led to the development of navigation. The skill of navigation started as local knowledge of the place a pilot frequently travelled. While a local pilot was useful, and the world's major ports still employ harbour pilots today, general methods of navigation applicable to circumstances that could not be known in advance were needed as ships sailed into unknown waters. The skill of navigation was turned into abstract ideas about position in space and time.

The various ancient empires of the four river systems mastered all the skills of observation, record-keeping, measurement, and mathematics that would form the foundation of natural philosophy. While historians have increasingly acknowledged the intellectual debt we owe these civilizations, we do not trace our scientific heritage to the Egyptians, Babylonians, Indians, or Chinese. Part of the reason for this is simply chauvinism. Science was largely a European creation, so there was a preference for beginning the heritage of natural philosophy with European sources rather than African or Asian ones.

There is, however, a more profound reason to start natural philosophy with the Greeks rather than the older cultures, despite their many accomplishments. Although these older cultures had technical knowledge, keen observational skills, and vast resources of material and information, they failed to create natural philosophy because they did not separate the natural world from the supernatural world. The religions of the old empires were predicated on the belief that the material world was controlled and inhabited by supernatural beings and forces, and that the reason for the behaviour of these supernatural forces was largely unknowable. Although there were many technical developments in the societies of the four river cultures, the intellectual heritage was dominated by the priests, and their interest in the material world was an extension of their concepts of theology. Many ancient civilizations, such as the Egyptian, Babylonian, and Aztec empires, expended a large proportion of social capital (covering such things as the time, wealth, skill, and public space of the society) on religious activity. The Great Pyramid, built as the tomb for the Pharaoh Khufu (also known as Cheops), rises 148 metres above the plain of Giza and is the largest of the pyramids. It is an astonishing engineering feat and tells us a great deal about the power and technical skills of the people who built it. But the pyramids also tell us about a society that was so concerned about death and the afterlife that its whole focus could be on the building of a giant tomb.

The very power of the four river centres may have worked against a change in intellectual activity. Social stratification and rigid class structure kept people in narrowly defined occupations. Great wealth meant little need to explore the world or seek material goods from elsewhere since the regions beyond the empire contained little of interest or value compared to what was already there. Although it was less true of the civilizations along the Indus-Ganges and Tigris-Euphrates river systems, which were more affected by political instability and invasions, both the Egyptian and Chinese civilizations developed incredibly complex societies with highly trained bureaucracies that grew increasingly insular and inward-looking.

The Greek World

It is impossible to be certain why the Greeks took a different route, but aspects of their life and culture offer some insight. The Greeks were not particularly well-off, especially when compared to their neighbours the Egyptians. Although unified by language and shared heritage, Greek society was not a single political entity but a collection of city-states scattered around the Aegean Sea and the eastern end of the Mediterranean. These city-states were in constant competition with each other in a frequently changing array of partnerships, alliances, and antagonisms. This struggle extended to many facets of life, so that it included not just trade or military competition but also athletic rivalry (highlighted by the athletic and religious festival of the Olympics); the pursuit of cultural superiority by claiming the best poets, playwrights, musicians, artists, and architects; and even intellectual competition as various city-states attracted great thinkers. This pressure to be the best was one of the spurs to exploration that allowed the Greeks to bring home the intellectual and material wealth of the people they encountered.

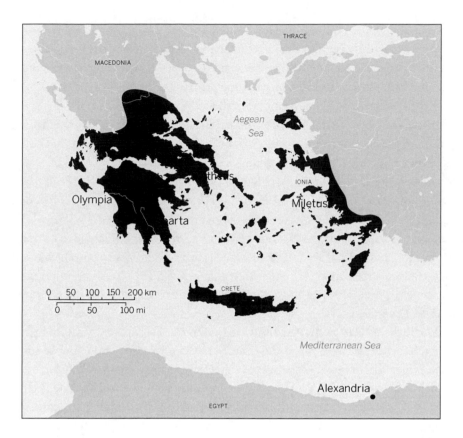

1.1 THE GREEK WORLD

Another factor was the degree to which Greek life was carried out in public. Much of Greek social structure revolved around the marketplace or *agora*. This was not just a place to shop but a constant public forum where political issues were discussed, various medical services were offered, philosophers debated and taught, and the news and material goods of the world was disseminated. The Greeks were a people who actively participated in the governance of the state and were accustomed to debate and discussion of matters of importance as part of the daily course of life. Greek law, while varying from state to state, was often based on the concept of proof rather than the exercise of authority. The public exchange of ideas and demand for individual say in the direction of their political and cultural life gave the Greeks a heritage of intellectual rigour and a tolerance for alternative philosophies. The vast range of governing styles that coexisted in the city-states, from tyranny to democracy, show us a willingness to try new methods of dealing with public issues.

Combined with the competitiveness of the Greeks, this meant that they were not only psychologically prepared to take on challenges but also accustomed to hearing and considering alternative views. They absorbed those things they found useful from neighbouring civilizations and turned them to their own needs.

Greek religion also differed from that of their neighbours. For the Greeks, the gods of the pantheon were much more human in their presentation and interaction with people. Mortals could argue with the gods, compete against them, and even defy them, at least for a time. Although the Greek world was still full of spirits, Greeks were less inclined to imbue every physical object with supernatural qualities. While there might be a god of the seas to whom sailors needed to make offerings, the sea itself was just water. The religious attitude of Greeks was also less fatalistic than that of their neighbours. While it might be impossible to escape fate, as the story of Oedipus Rex shows, it was also the case that the gods favoured those who helped themselves. At some fundamental level, the Greeks believed that they could be the best at everything, and they did not want to wait for the afterlife to gain their rewards.

Although there were many positive things about Greek society, we should also remember that the Greeks had the time and leisure for this kind of public life because a large proportion of the work to keep the society going was done by slaves. Although the conditions of slavery varied from city-state to city-state, even in democratic Athens (where democracy was limited to adult males of Athenian birth), most of the menial positions and even the artisan class were made up of slaves. Those who worked with their hands were at the bottom of the social hierarchy.

Thales to Parmenides: Theories of Matter, Number, and Change

Whether these elements of Greek society and social psychology are sufficient to explain why the Greeks began to separate the natural from the supernatural is difficult to prove. Yet this separation became a central tenet for a line of philosophers who began to appear in Ionia around the sixth century BCE. The most famous of these was Thales of Miletus (c. 624–c. 548 BCE). We know very little about Thales or his work. Most of what comes down to us is in the form of comments by later philosophers. He was thought to have been a merchant, or at least a traveller, who visited Egypt and Mesopotamia where he was supposed to have learned geometry and astronomy. Thales argued that water was the prime constituent of nature and that all matter was made of water in one of three forms: water, earth, and mist. He seems to be borrowing from the material conception of the Egyptians, who also considered earth, water, and air to be the primary constituents of the material world, but he took it one step further by starting with one element. Thales pictured the world as a sphere (although it might have been drum-shaped) that floated on a celestial sea.

Even in this fragmentary record of Thales' philosophy, two things stand out. First, nature is completely material; there are no hints of supernatural constituent elements. This does not mean that Thales discarded the gods but rather that he thought that the universe had a material existence independent of supernatural beings. The second point is that nature functions of its own accord, not by supernatural intervention. It follows that there are general or universal conditions governing nature and that those conditions are open to human investigation and understanding.

Following Thales was his student and disciple Anaximander (c. 610–c. 545 BCE). Anaximander added fire to the initial three elements and produced a cosmology based on the Earth at the centre of three rings of fire. These rings were hidden from view by a perpetual mist, but apertures in the mist allowed their light to shine through, producing the image of stars, the sun, and the moon. Like Thales, Anaximander used a mechanical explanation to account for the effects observed in nature. His system presented some problems since it placed the ring of fire for the stars inside the rings of fire for the moon and the sun. He may have addressed these issues elsewhere, but that information is lost to us.

Anaximander also tried to provide a unified and natural system to account for animal life. He argued that animals were generated from wet earth that was acted upon by the heat of the sun. This placed all four elements together as a prerequisite for life. This conception of spontaneous generation was borrowed from earlier thinkers and was likely based on the observation of events such as the appearance of insects or even frogs from out of the ground. Anaximander took the theory a step further by arguing that simpler creatures changed into more complex ones. Thus, humans were created from some other creature, probably some kind of fish. This linked the elements of nature with natural processes rather than supernatural intervention to create the world that we see.

The Ionian concern with primary materials and natural processes would become one of the central axioms of Greek natural philosophy, but by itself it was insufficient for a complete philosophical system. At about the time

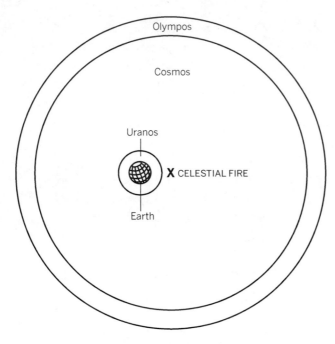

1.2 THE UNIVERSE ACCORDING TO PYTHAGORAS

Anaximander was working on his material philosophy, another group of Greeks was developing a conception of the world based not on matter but on number. This thread of philosophy comes down to us from Pythagoras (c. 582–500 BCE). It is unclear if there actually was a single historical figure named Pythagoras. Traditionally, he was thought to have been born on the island of Samos and to have studied Ionian philosophy, perhaps even as a student of Anaximander. He was supposed to have threatened the authority of the tyrant Polycrates on Samos and was forced to flee the island for Magna Graecia (Italy).

Because Pythagoras's followers became involved in conflicts with local governments, the Pythagoreans should not be regarded as simply a wandering band of mathematicians. Their lives were based, in fact, on a religion full of rituals. They believed in immortality and the transmigration of souls, but at the heart of Pythagoreanism was the conception of the universe based on number. All aspects of life could be expressed in the form of numbers, proportions, geometry, and ratios. Marriage, for example, was given the number five as the union of the number three representing man and the number two representing woman. Although there were mystical aspects of the number system, the Pythagoreans attempted to use mathematics to quantify nature. A good example can be seen in their demonstration of musical harmony. They showed that the length of a string determined the note produced, and that note was then related exactly to other notes by fixed ratios of string length.

The Pythagoreans developed a cosmology that divided the universe into three spheres. (See figure 1.2.) Uranos, the least perfect, was the sublunar realm or terrestrial sphere. The outer sphere was Olympos, a perfect realm and the home of the gods. Between these two was Cosmos, the sphere of moving bodies. Since it was governed by the perfection of spheres and circles, it followed that the planets and fixed stars moved with perfect circular motion. The word "planet" comes from the Greek for "wanderer," and it was used to identify these spots of light that constantly moved and changed position against the fixed stars and relative to each other. The planets were the Moon, Sun, Mercury, Venus, Mars, Jupiter, and Saturn. The fixed stars orbited without changing their position relative to each other, and it was from these that the constellations were formed.

While this arrangement was theologically satisfying, it led to one of the most perplexing problems of Greek astronomy. The philosophy of perfect circular motion did not match observation. If the planets were orbiting the Earth at the centre of the three-sphere universe, they should demonstrate uniform motion—and they did not. To resolve this problem, the Pythagoreans moved the Earth out of the centre of the sphere and created a point—home to a celestial fire—that was the centre of uniform motion. This kept the Earth motionless and resolved the issue of the observed variation in the velocity and motion of the planets. The desire to keep the Earth at the centre of the universe and preserve the perfection of circular motion led most of the later Greek philosophers to reject the Pythagorean solution. A radical solution to this problem was proposed by Aristarchus of Samos (c. 310–230 BCE), who argued for a heliocentric (sun-centred) model, but his ideas gained little support because they not only violated common experience but ran against religious and philosophical authority on the issue.

One of the most famous geometric relations comes down to us from the Pythagoreans, although they did not create it. This is the "Pythagorean theorem" that relates the length of the hypotenuse of a triangle to its sides. This relationship was well known to the Egyptians and the Babylonians and probably came from surveying and construction. The relationship can be used in a handy instrument by taking a rope loop marked in 12 equal divisions that when pulled tight at the 1, 4, and 8 marks produces a 3–4–5 triangle and a 90° corner. (See figure 1.3.) The Pythagoreans used geometric proof to demonstrate the underlying principle of this relationship.

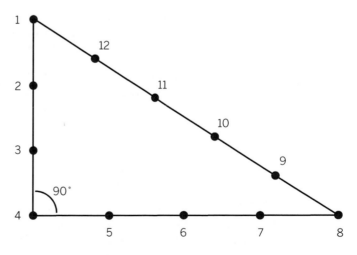

1.3 USING THE PYTHAGOREAN RELATION TO CREATE A RIGHT ANGLE

A rope with 12 evenly spaced knots when pulled at 1, 4, and 8 creates a right angle at 4. This simple device was known to the Egyptians and used for surveying and building.

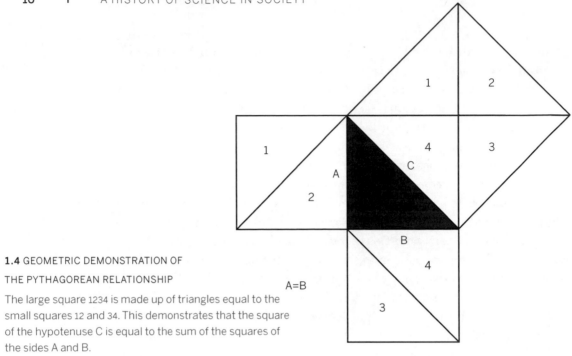

1.4 GEOMETRIC DEMONSTRATION OF
THE PYTHAGOREAN RELATIONSHIP
The large square 1234 is made up of triangles equal to the
small squares 12 and 34. This demonstrates that the square
of the hypotenuse C is equal to the sum of the squares of
the sides A and B.

Despite the mystical aspects of a world composed of number, the foundation
of Pythagorean thought places the essential aspects of natural phenomena within
the objects themselves. In other words, the world works the way it does because
of the intrinsic nature of the objects in the world and not through the intervention of unknowable supernatural agents. Ideal forms, especially geometric objects
such as circles and spheres, existed as the hidden superstructure of the universe,
but they could be revealed, and they were not capriciously created or changed by
the gods.

The degree to which the Pythagoreans desired a consistent and intrinsically
driven nature can be seen in the problem created by "incommensurability," referring
to things that had no common measure or could not be expressed as whole number
proportions such as 2:3 or 4:1. The Pythagoreans argued that all nature could be
represented by proportions and ratios that could be reduced to whole-number
relationships, but certain relationships cannot be expressed this way. In particular,
the relationship between the diagonal and the side of a square cannot be expressed
as a ratio of integers such as 1:2 or 3:7. As figure 1.4 demonstrates, the relationship
can be shown geometrically, but the arithmetic answer was not philosophically
acceptable since it required a ratio of 1:$\sqrt{2}$, which could not be expressed as an

integer relation. No squared number could be subdivided into two equal square numbers, nor in the case of √2 can the number be completely calculated.[1] According to legend, the Pythagorean Hippapus, who discovered the problem, was thrown off the side of a ship by Pythagoras to keep incommensurability secret.

The problems of Greek mathematics were compounded by two practical issues. The Greeks did not use a decimal or place-holder system of arithmetic but used letters to represent numbers. This made calculations and more complex forms of mathematics difficult. In addition, even though the Greeks and the Pythagoreans in particular were extremely powerful geometers, they did not have a system of algebra, and proofs were not based on "solving for unknowns." Geometric proofs were created to avoid unknown quantities. These two aspects of Greek mathematics put limits on the range of problems that could be addressed and probably encouraged their concentration on geometry.

While the Ionians investigated the material structure of the world and the Pythagoreans concentrated on the mathematical and geometric forms, another aspect of nature was being investigated by Greek thinkers. This was the issue of change. Motion, growth, decay, and even thought are aspects of nature that are neither matter nor form. No philosophy of nature could be complete without an explanation of the phenomena of change. At the two extremes of the issue were Heraclitus of Ephesus (c. 550–475 BCE) and Parmenides of Elea (fl. 480 BCE). Heraclitus argued that all was change and that nature was in a constant state of flux, while Parmenides asserted that change was an illusion.

Heraclitus based his philosophy on a world that contained a kind of dynamic equilibrium of forces that were constantly struggling against each other. Fire, at the heart of the system and the great image of change for Heraclitus, battled water and earth, each trying to destroy the others. In a land of islands, water, and volcanoes, this had a certain pragmatic foundation. Heraclitus's most famous argument for change was the declaration that you cannot step into the same river twice. Each moment, the river is different in composition as the water rushes past, but, in a more profound sense, you are as changed as the river and only the continuity of thought gives the illusion of constancy.

For Parmenides, change was an illusion. He argued that change was impossible since it would require something to arise from nothing or for being to become non-being. Since it was logically impossible for nothing to contain something

....

1. Like π, √2 is part of a collection of numbers that were later called "irrational," because they do not form proper ratios.

(otherwise it would not have been nothing in the first place), there could be no mechanism to change the state of the world.

Parmenides' best-known pupil, Zeno (fl. 450 BCE), presented a famous proof against the possibility of motion. His proof, called Zeno's paradox, comes in a number of forms but essentially argues that to reach a point, you must first cover half the distance to the point. To get to that halfway point, you would first need to cover half the distance (i.e., one-quarter of the full distance), and therefore one-eighth, one-sixteenth, and so on. Since there are an infinite number of halfway points between any two end points, it would take infinite time to cover the whole distance, making it impossible to move. (See figure 1.5.)

Our modern perception seems to favour Heraclitus over Parmenides, but they share a common concern. Each philosopher was attempting to establish a method for understanding the events in the world based on the intrinsic or natural action of the world. They were also attempting, as the Ionians and the Pythagoreans did, to establish a method for determining what certain knowledge was. Statements about the condition of the world had to be supported by a proof that could be examined by others and did not rely on special knowledge. They were asking epistemological questions, that is, questions about how someone could come to know something and just what that "something" could be. The Greek natural philosophers did not frame their questions as inquiries into the behaviour of gods or supernatural agents but rather asked such questions as: What in the world around us is fundamental and what is secondary? What system (outside revelation) can a thinker use to determine what is true and what is false? To what degree should the senses be trusted?

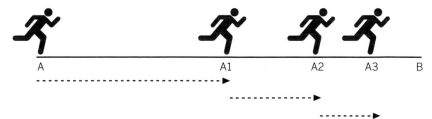

1.5 ZENO'S PARADOX

As the runner covers half the distance from "A" to "B," he must first cover the distance from "A1" to "B," then half the distance from "A2" to "B," and so on. Since there are an infinite number of halfway points, and it takes a finite amount of time to move from point to point (even though the time to cover the distance is very small), it will thus take an infinite amount of time to get from "A" to "B."

For Parmenides, the senses were completely untrustworthy and only logic could produce true or certain knowledge. Heraclitus at first seemed to have more faith in the senses, but in fact he reached a very similar conclusion. Any appearance of stasis, even something as simple as one rock resting on another, is an illusion, and only logic can be relied upon to make clear what is actually happening in nature.

Socrates, Plato, Aristotle, and the Epicureans: The Ideal and the Real

The philosophical threads of Thales, Pythagoras, Heraclitus, Parmenides, and many others came together in the work of the most powerful group of Greek thinkers, who were at the intellectual hub of Athens in the fifth century BCE. Socrates (470–399 BCE) established a context for natural philosophy by completely rejecting the study of nature as being largely unworthy of the philosopher's thought and by creating the image of selfless dedication to the truth that helped form the image of the "true" intellectual to this very day. Socrates' rejection of the study of nature mirrored the increasing disdain the intellectual elite felt for the merchant and craft class and their material concerns. Philosophy was supposed to be above the petty concerns of the day-to-day world, and philosophers were not, both literally and figuratively, to get their hands dirty.

For Socrates, the real world was the realm of the Ideal. Since nothing in the material world could be perfect, it followed that the material world must be secondary to the ideal. For example, while one could identify a beautiful person, the concept of beauty must have been present prior to the observation or we would be unable to recognize the person as beautiful. Further, while any particular beautiful material thing must necessarily fade and decay, the concept of beauty continues. It thus transcends the material world and is eternal.

This idealism also applied to the comprehension of the structure of the material world. Any actual tree was recognizable as a tree only because it reflected (imperfectly) the essence of "tree-ness," or the form of the ideal tree. These ideal forms were available to the human intellect because humans had a soul that connected them to the perfect realm. Socrates believed that, because of this, we actually had within ourselves the knowledge to understand how things worked. With skillful questions, this innate knowledge could be revealed, and from this process we get the Socratic method, a form of teaching based not on the instructor giving information to the student but asking a series of questions that guides the student's thoughts to the correct understanding of a topic.

Socrates' philosophy led him to question everything, including the government of Athens. He was convicted of corrupting the city's youth, but rather than asking for exile, he chose death. He drank a potion of the poison hemlock, with the firm belief that he was leaving the imperfect, corrupt material world for the perfection of the Ideal realm.

Socrates left no written material, so what we know of his teachings largely comes to us from his most famous pupil, Plato (427–347 BCE). The son of an aristocratic Athenian family, Plato wrote a series of dialogues based on Socrates' ideas and likely drawn from actual discussions. Although Plato's later work shifted away from its Socratic roots, he preserved the general premise of Ideal forms. One of Plato's other teachers was Theodorus of Cyrene, a Pythagorean, who taught him the importance of mathematical idealism. Although Plato accepted the primacy of the Ideal, he did not go as far as Socrates in his rejection of the material world.

Plato's primary interests were ethical and political. In his most famous work, *The Republic*, he explored what he considered ideal society and the problems of social organization. He did introduce natural philosophy, but it was in a lower realm of consideration and used mostly as a tool for consideration of the underlying structure of the cosmos. In the allegory of the cave, found in Book VII of *The Republic*, Plato argued that people are like prisoners in a dark cave who, from childhood, see only a strange kind of shadow play. Because the prisoners have no other reference, the shadows are taken to be reality. To see reality, the prisoners must free themselves and look upon the real world under the light of the sun. In this story, Plato argued that what we perceive through our senses is an illusion, but logic and philosophy can reveal the truth. The material world was explored in more detail in his *Timaeus*, where he presented a system of the four terrestrial elements of earth, water, air, and fire. The supralunar or celestial realm was made of a perfect substance, the ether, and was governed by a different set of physical rules. This system gained general acceptance among Greek philosophers and became one of the axioms of natural philosophy.

Plato, unlike his teacher Socrates, was not content to espouse his philosophy in the *agora*. The solution to the problems of society was education, which meant training students in a philosophy based on logic and a pursuit of knowledge of the Ideal. To this end, Plato founded a school in 385 BCE. Constructed on land once owed by the Athenian hero Academos, it became known as the Academy. It did

CONNECTIONS

Natural Philosophy and Patronage: Aristotle and Alexander the Great

The relationship between patron and client has been an important part of the development of natural philosophy and science from the time of the Ancient Greeks. Aristotle was heavily influenced by the materials he received from Alexander the Great, and his fame spread even farther because of the king's patronage.

In 343 BCE, King Philip II of Macedon asked Aristotle to join his court as the tutor to his son Alexander. Aristotle's father had been Philip's personal physician, so there was already a connection between Aristotle and Philip's family. The call to go to Macedon came at a time when Aristotle was pursuing biological and philosophical research on his own because he had quit his teaching position at the Academy, the school established by Plato in Athens.

Aristotle remained at court for seven years, teaching the sons of Macedonian nobles. Aristotle found Alexander a good, if somewhat mercurial, student who wanted to be the best at whatever he did. When Philip was assassinated in 336 BCE, Alexander became the king and went on to conquer Greece and then most of the known world, including Asia Minor, Egypt, and Persia. He remained close friends with Aristotle, corresponding with his teacher throughout his life. He also sent Aristotle hundreds of samples of plants and animals, and over 10,000 scrolls from distant lands.

In 334 BCE Aristotle returned to Athens and established a new school called the Lyceum. Under the patronage of Alexander, the school thrived and Aristotle wrote a number of his most important works in this period, including *Physics*, *Parts of Animals*, and *De Anima*. The vast library created from Alexander's gifts helped Aristotle with his philosophical work, while the plant and animal samples helped him with his biological research. Aristotle described fish, for example, that were not noted again in Europe for hundreds of years, and developed a robust classification system because of this wide experience.

Alexander was a philosopher-king: literate, well-educated, and curious about more than just the necessities of warfare and politics. His relationship with Aristotle became a model of patronage that many later natural philosophers from Alcuin to Descartes hoped to find for themselves.

not have the formal structure of modern schools, but in many ways it was the foundation for the concept of higher education. Students who had already been tutored in the basic principles of subjects such as rhetoric and geometry travelled to the Academy to engage in discussion and debate under the auspices of a more senior philosopher in a kind of seminar atmosphere.

Plato's most famous student was Aristotle (384–322 BCE). A brilliant thinker, Aristotle had expected to become the head of the Academy when Plato died, but this position was denied him, going instead to Plato's cousin Speusippas, of whom little is known. Disappointed at having been passed over, Aristotle left Athens and travelled north. In 343 BCE he became the tutor to Alexander, son of Philip II, King of Macedon. When Philip died, Alexander became the leader of the Macedonians and proceeded to unify (that is, conquer) all of Greece. Once that was accomplished, he set out to conquer the rest of the world. With the patronage of Alexander the Great, Aristotle returned to Athens and founded a rival school, the Lyceum, in 334 BCE. It was sometimes called the peripatetic school because the instructors and scholars did their work while walking around the neighbourhood.

Aristotle did not reject all of Plato's philosophy, sharing a belief in the necessity of logic and some aspects of Platonic Idealism. He was, however, far more interested in the material world. Although he agreed with Plato that the world was impure and our senses fallible, he argued that they were actually all we had. Our intellect could be applied only to what we observed of the world around us. With this as a basis, Aristotle set out to create a complete system of natural philosophy. It was a powerful and extremely successful project.

At the heart of Aristotle's system were two fundamental ideas. The first was a system to provide a complete description of natural objects. The second was a system to verify knowledge that would satisfy the demands of proof necessary to convince people who lived in a competitive, even litigious, society. The combination of these two components produced the apex of Greek natural philosophy. No aspect of Aristotle's philosophy depended on supernatural intervention, and only one entity, the unmoved mover, existed outside the system of intrinsic or natural action.

The first step in the description of natural objects was identification and classification. Aristotle was a supreme classifier. Much of his work was on biology, and among other things he grouped what we call reptiles, amphibians, and mammals by their characteristics, even grouping dolphins with humans. He also observed the development of chicks in hen eggs and tried to make sense of sexual reproduction.

As astute as many of his observations were, Aristotle saw them as an examination of a level of superficial distinction; it was the job of the philosopher to look beyond these secondary characteristics and seek the underlying structure of nature. To do this, it was necessary to determine what aspects of nature could not be reduced to simpler components. The simplest material components were the four elements, and all material objects in the terrestrial realm were composed of these four substances. The superficial distinction between objects was the result of the different proportions and quantities of the elements that made up the objects in the world.

The elements by themselves were not sufficient to account for the organization and behaviour of matter. Matter also seemed to have four irreducible qualities, which Aristotle identified as hot/cool and wet/dry. These were always present as pairs (hot/wet, cool/wet, hot/dry, cool/dry) in all matter, but were separate from the material. A loose analogy would be to compare the bounce of a basketball and a bowling ball. The degree of bounce of a basketball and a bowling ball are very different and depend on the material that each is made of, but the "bounciness" of the two balls can be studied separately from the study of the materials that compose the two types of ball.

While the four elements and the four qualities could describe the matter and quality of composed things, they did not explain how a thing came to be. For this, Aristotle identified four causes: formal, material, efficient, and final. The formal cause of a thing was the plan or model, while the material cause was the "stuff" used to create the object. The efficient cause was the agent that caused the object to come into being, and the final cause was the purpose or necessary condition that led to the object's creation.

Consider a stone wall around a garden. The formal cause of the wall is its plans and drawings. Without a plan detailing dimensions, it is impossible to know how much stone will be required to build it. The material cause of the wall is the stones and mortar. These materials impose certain restrictions on the finished wall; it might be possible to draw a plan for a 30-metre high wall with a base only 20 centimetres wide, but such a wall cannot be constructed in reality. The efficient cause is the stonemason; again, certain restrictions will be imposed on the wall by the limits of the mason's abilities. The final cause is the reason to build the wall—to keep the neighbour's goat out of our garden, for example.

Although Aristotle and Plato's conception of the four elements could be reduced to a kind of particle model with a geometric structure (fire, for example, was composed of triangles), in general they treated the elements as a continuous substance. This view was challenged by the Epicureans, who proposed an even more materialistic model of nature. The philosopher Epicurus (342–271 BCE), like Plato, was from an aristocratic Athenian family. He founded a philosophical school known as the Garden and revived the work of an earlier philosopher, Democritus (c. 460–c. 370 BCE). Democritus had argued for a materialistic understanding of the universe, and the Epicureans pictured the world as constructed of an innumerable (but not infinite) number of atoms that were indestructible. The appearance and behaviour of matter were based on the varying size, shape, and position of the particles.

Epicurean natural philosophy was the most mechanistic Greek philosophy. In addition to challenging the material foundation of nature, the Epicureans also challenged the path to knowledge of nature, arguing that knowledge could only come from the senses. Because knowledge of nature did not require the intellectual refinement of logic or mathematics, it was knowledge open to all, not just learned men. This belief in knowledge from the senses contributed to the reputation of the Epicureans as sensualists, which did not help the philosophy when it was attacked as atheistic and decadent by Jewish, Islamic, and Christian scholars in later years. Although there was suspicion of all Greek philosophy by later theological thinkers, Aristotle's system was more easily revised than the Epicurean because it ultimately depended on axioms that could be ascribed to God. Thus, Epicurean thought was largely condemned or ignored until the seventeenth century when it gained a titular place as the foundation of modern studies of matter because of its proto-atomic model. Thus, it is seen as the ancient precursor to modern chemistry.

Aristotelian Theories of Change and Motion

The three fundamental aspects of matter (elements, qualities, and causes) in the Aristotelian system cannot assemble themselves into the universe; to bring everything together there must be change and motion. There are two kinds of motion. The first, natural motion, is an inherent property of matter. In the terrestrial realm all elements have a natural sphere, and they attempt to return to their natural place by moving in a straight line. However, because many objects in the world are mixtures of the four elements, natural motion is restrained in various ways. A tree, for example, contains all four elements in some proportion, but it grows a certain way with the roots going down because the earth element wants to go down while the crown grows up as its air and fire elements try to go up.

Plato and Aristotle accepted the Pythagorean idea that the matter in the celestial realm was perfect and that its inherent natural motion was also perfect, travelling in a uniform and immutable circle, which was the perfect geometric figure. Aristotelian astronomy thus required the objects in space to move according to this dictum. While this was a reasonable assumption for most of the objects that could be observed, such as the sun, moon, and stars, it created problems for later astronomers. (See figure 1.6.)

Other forms of motion, particularly locomotion, required motion to be introduced to the universe. For this, Aristotle traced a chain of motion back from observation to origin. Anything moving had a mover, but that mover had to have something moving it, and so on. Take as an example an archer shooting an arrow. We see an arrow fly through the air, and we can observe that it was the bow moving that moved the arrow. The archer

Schema huius praemissae diuisionis Sphaerarum

1.6 THE ARISTOTELIAN COSMOS

makes the bow move by the motion of muscles, and the muscles are made to move by the will of the archer. The mind thinks (which is a kind of motion as well) because of a soul, and the body exists because it was the product of the athlete's parents. Birth and growth are also forms of motion. The archer's parents were created by the grandparents, and so on. To prevent this from becoming a completely infinite regress, there has to be some point at which a thing was moved without being moved itself by some prior thing. This is the unmoved mover. In a sense, the unmoved mover kick-started motion in the universe by starting the great chain of action by a single act of will.

Let us return to the arrow as it flies along. As long as the bow is in contact with it, we can see that it is the bow and the muscles that are making it move, but what keeps it moving after it has left the bowstring? The aspect of its motion toward the ground is covered by its natural place as the heavy earth element of the arrow attempts to return to its proper sphere. The continuation of motion,

1.7 THE ARROW'S MOTION
ACCORDING TO ARISTOTLE

The arrow interacts with the air as it moves to continue its "unnatural" motion. This system may seem awkward, but it was likely based on observation of motion through water. An oar pulled through water seems to compress the water (it clearly mounds up) on the front surface, while eddies and voids seem to form around the back surface of the oar. The water in front then rushes around the oar to fill in the space at the back.

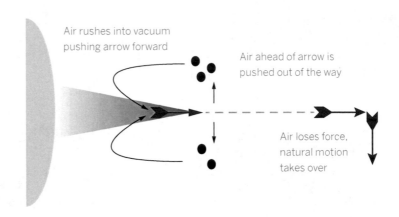

Air rushes into vacuum pushing arrow forward

Air ahead of arrow is pushed out of the way

Air loses force, natural motion takes over

Aristotle reasoned, had to have something to do with motion being added to the object as it moves. He concluded that the arrow was being bumped along by its very passage through the air. The arrow was pushing the air out of its natural place, in effect compressing it at the front and creating a rarefied or empty area at the back. The air rushed around the arrow to restore the natural balance and, in doing so, bumped the arrow ahead. Since the air resisted being moved from its natural place, it would eventually stop the forward flight of the arrow. (See figure 1.7.)

It also followed from Aristotle's system that the amount of element in an object governed its rate of motion. An arrow, constructed of wood and thus not containing a large amount of earth element, would stay in motion over the ground longer than a rock composed almost completely of earth element. This led Aristotelians to argue that if a small rock and a large rock weighing ten times as much were dropped together, the large rock would fall ten times faster than the small rock.

Aristotelian Logic

While understanding the structure of matter and motion was important, such knowledge was not by itself sufficient to understand the world. This was, in part, because the senses could be fooled and were not entirely accurate, but it was also because observation was confined to the exterior world and could not by itself reveal the underlying rules or structure that governed nature. That could be discovered only by the application of the intellect, and that meant logic. While Aristotle returned to the subject of logic repeatedly, his logic was most clearly presented in his two works on the subject, the *Posterior Analytics* and the *Prior Analytics*. At the heart of his logical system was the syllogism, which offered a method to prove a relationship and thereby produce reliable or certain knowledge. We continue to use syllogistic logic today as a method of verifying the reliability of statements. One of the most famous syllogisms says:

1. All men are mortal.   *Major premise, derived from axioms or previously*
                             established true statements.
2. Socrates is a man.   *Minor premise. This is the condition being investigated.*
3. Therefore, Socrates is mortal.   *Conclusion, which is deduced from the premises.*

The syllogism was a powerful tool to determine logical continuity, but it could not by itself reveal whether a statement is true, since false but logical syllogisms can be constructed.

1. All dogs have three legs.
2. Lassie has four legs.
3. Therefore, Lassie is not a dog.

The second syllogism is as consistent as the first, but because the major premise is false, the conclusion is false. The axiom "dogs have three legs" does not stand the test of observation or definition, and so the syllogism fails. Thus, it is not surprising that Greek philosophers expended a great deal of effort on the discovery and establishment of axioms. Axioms were irreducible, self-evident truths. They represented conditions that must exist if the world was to function, but recognizing them was difficult. Aristotle concluded that axioms could be recognized only by the agreement of all learned men, which echoed Greek political discourse. An example of an axiom is the operation of addition, which must be accepted as a necessary mathematical operation or all of arithmetic collapses. The property of addition cannot be broken down into simpler operations; multiplication, on the other hand, can be broken down into repeated addition and is thus not axiomatic.

The problem of what was axiomatic and how to be sure of axiomatic statements was at the centre of debates over natural philosophy and science, in part because the axioms of previous generations often became the target of investigation and reduction for new thinkers. The philosophical and practical attacks on axioms at times made some scholars unsure whether any knowledge was reliable, while it set others, such as René Descartes (1596–1650), on a search for a new foundation of certainty.

The power of Aristotle's system was its breadth and completeness. It integrated the ideas that had been developed and philosophically tested, in some cases for several hundred years, with his own observations and work on logic. It presented a system for understanding the world that was almost completely intrinsically derived. With the exception of the unmoved mover, no part of his system required supernatural intervention to function, and further, it was based on the belief that all of nature could be understood. The comprehensibility of nature became one of the characteristics of natural philosophy that separated it from other studies such as theology or metaphysics.

Aristotle's system was a masterful use of observation and logic, but it did not include experimentation. Aristotle understood the concept of testing things, but he rejected or viewed with distrust knowledge gained by testing nature, because such tests only showed how the thing being tested acted in the test rather than in nature. Since testing was an unnatural condition, it was not part of the method of natural philosophy, which was to understand things in their natural state. It is tempting to find fault with Aristotle because of his rejection of experimentation, but this would be to argue that Aristotle's objectives must have been the same as those of modern science. The object of study for Aristotle and modern science was nature and how nature functions, but the forms of the questions asked about nature were very different. One of the central questions for Aristotle and other natural philosophers was teleological, asking "To what end does nature work?" They assumed that only through observation and logic could this question be answered.

Euclid and the Alexandrians

After the death of Aristotle, both the Academy and the Lyceum continued to be major centres for philosophical education, but the heart of Greek scholarship began to shift to Alexandria. This movement was spurred after 307 BCE when the ruler of Egypt, Ptolemy I (who had been one of Alexander's generals) invited Demetrius Phaleron, the deposed dictator of Athens, to move to his capital at Alexandria. Alexandria was an ideal location as a trade hub that linked Africa, Europe, the Middle East, and Asia. Demetrius was credited with advising Ptolemy to establish a collection of texts and establish a temple to the Muses, who were the patrons of the arts and sciences. Although its exact founding and early history are unclear, the temple to the Muses became the Museum, from which our modern use of the term descends. Part of the Museum was the library, which became increasingly important and eventually overshadowed the Museum in historical recollection. The Great Library of Alexandria eventually housed the greatest collection of Greek texts and was the chief repository and education centre for Aristotelian studies after the decline of Athens.

One of the great figures to be associated with the Museum was Euclid (c. 325–c. 265 BCE).[2] His most enduring work was the *Elements*, a monumental compilation of mathematical knowledge that filled 13 volumes. While the majority of the material in the *Elements* was a recapitulation of earlier works by other scholars, two factors raised it above a kind of mathematical encyclopedia. The first was the systematic presentation of proofs, so that each statement was based on a logical demonstration of what came before. This not only gave the mathematical proofs reliability but also influenced the method of presenting mathematical and philosophical ideas to the present day. These proofs were based on a set of axioms such as the statement that parallel lines cannot intersect or that the four angles created by the intersection of two lines are two pairs of equal angles and always equal 360° in total.

The second factor was the scope of the work. By bringing together the foundation of all mathematics known to the Greeks, the *Elements* was a valuable resource for scholars and became an important educational text. It covered geometric definitions and construction of two- and three-dimensional geometric figures, arithmetic operations, proportions, number theory including irrational numbers, and solid geometry including conic sections. In a time when all manuscripts had to be copied by hand, the *Elements* became one of the most widely distributed and widely known texts.

Greek natural philosophy was most notable for its philosophical systems, but those systems should not be seen as being removed from the real world or as some kind of irrelevant intellectual pastime. One of the purposes of Aristotelian natural philosophy was to make the world known, and a known world was a classified and measured world. Eratosthenes of Cyrene (c. 273–c. 192 BCE) set out to measure the world. Eratosthenes was a famous polymath who worked in many fields, especially

---

2. Like Pythagoras, there is some dispute as to whether Euclid was a real person or a name applied to a collective of scholars. From later commentators and internal evidence, Euclid may have been educated in Athens, perhaps at Plato's Academy, and then moved to Alexandria.

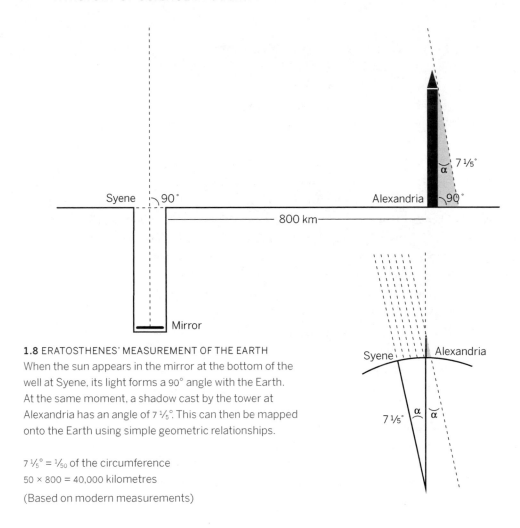

1.8 ERATOSTHENES' MEASUREMENT OF THE EARTH
When the sun appears in the mirror at the bottom of the
well at Syene, its light forms a 90° angle with the Earth.
At the same moment, a shadow cast by the tower at
Alexandria has an angle of 7 ⅕°. This can then be mapped
onto the Earth using simple geometric relationships.

7 ⅕° = ¹⁄₅₀ of the circumference
50 × 800 = 40,000 kilometres
(Based on modern measurements)

mathematics, and who became the chief librarian of the Museum in Alexandria
about 240 BCE. He applied his concepts of mathematics to geography and came up
with a method to measure the circumference of the Earth. That the Earth was a
sphere was long understood by the Greeks and was taken as axiomatic in
Aristotle's philosophy, but an accurate measurement was a challenge. Eratosthenes
reasoned that by measuring the difference in the angle of a shadow cast at two
different latitudes at the same moment, he could calculate the circumference. By
knowing the angle formed by the two lines radiating from the centre of the Earth
to the measuring points and the distance between the two points at the surface,
he was able to determine the proportion of the globe that distance represented.

(See figure 1.8.) From this, it was a simple matter to work out the circumference of the whole globe. His answer was 250,000 *stadia*. There has long been an argument about just how accurate this measurement was, since it is not clear what length of stadia Eratosthenes was using, but it works out to about 46,250 kilometres, which is close to the current measurement of 40,075 kilometres at the equator.

Archimedes, the Image of the Philosopher

The intellectual heritage of the Greeks, particularly that of Aristotle and Plato, was profound, but it was not solely their thought that they contributed. The Greeks also helped to create the image of the philosopher, an image that persists in various forms to the present day. Long before students have learned enough to comprehend the complex ideas of the philosophers, they have been exposed to the image. Even more famous than Socrates accepting death, the story of Archimedes (c. 287–212 BCE) has shaped the cultural view of philosophers.

Archimedes lived most of his life in Syracuse. He may have travelled to Alexandria and studied with Euclidean teachers at the Museum; it is clear that later in his career he knew and corresponded with mathematicians there. Among his accomplishments Archimedes determined a number for pi—relating the circumference, diameter, and area of a circle—and then extended this work to spheres. He established the study of hydrostatics, investigating the displacement of fluids, asking why things float, and the relationship between displaced fluids and weight. This has come down to us as Archimedes' principle that a body immersed in a fluid is buoyed up by a force equal to the weight of the fluid displaced by the body. Archimedes also determined the laws of levers through geometric proof.

As powerful as Archimedes' mathematics and philosophical work might have been, it was the legends that grew up around him that made him a memorable figure. His work was not confined to intellectual research, since he also created mechanical devices. Chief among these were the war machines he built to help defend Syracuse from the Romans during the Second Punic War. These included various ballistic weapons and machines to repel ships from docking. Although Archimedes did not invent Archimedes' screw (which consists of a rotating spiral tube used to lift water), his name was attached to it as the kind of thing he would have invented.

The famous story about Archimedes inventing burning mirrors or using polished shields to set fire to Roman ships using the reflected light of the sun was a myth created long after his death. Although theoretically possible, most modern recreations of the burning mirrors have shown that it would have been at best impractical, requiring the Roman ships to remain still for a significant period, and having no Roman notice the fire until it was large enough to have done significant damage.

Archimedes in the bath is the best-known tale from the philosopher's life. Hiero, the king of Syracuse, was concerned that the gold he had given craftsmen to make a crown had been adulterated with less valuable metal, but once the crown was made, how could the fraud be detected? Archimedes was supposed to have solved the problem while in the public baths when he realized that it was a hydrostatic problem. The gold would displace less water than a similar weight of silver because the gold was denser. He leapt from the bath and ran naked through the city, exclaiming "Eureka!" meaning "I have found it." No historical record exists that this happened, and it would have been difficult to use the displacement method with the tools available to Archimedes, but he could easily have solved this problem using a hydrostatic balance, a device that he wrote about and used.

Archimedes' death also became the stuff of legend. Plutarch (45–120 CE) tells the story in *Plutarch's Lives*:

> Archimedes, who was then, as fate would have it, intent upon working out some problem by a diagram, and having fixed in his mind alike and his eyes upon the subject of his speculation, he never noticed the incursion of the Romans, nor that the city had been taken. In this transport of study and contemplation, a soldier, unexpectedly coming up to him, commanded him to follow to Marchellus; which he declining to do before he had worked out his problem to a demonstration, the soldier, enraged, drew his sword and ran him through.[3]

Whether the legends are based on actual events is less important than the image of the ideal scholar they have come to represent. While the historical image of Archimedes has ranged from absent-minded philosopher to man of action to the "Divine Archimedes" as Galileo called him, the image of the true philosopher is that of a person above mundane concerns or personal self-interest. He is selfless, absorbed in study to the exclusion of all else, and perhaps a touch socially unaware. While Archimedes made mechanical devices and thus has also been associated with engineers, he was far more interested in philosophy than such contrivances. He

---

3. Plutarch, *Plutarch's Lives*, trans. John Dryden (New York: Random House, 1932) 380.

became the exemplar of a good scientist who can turn his hand to both theoretical and practical projects. While Aristotle and Plato can be revered as great intellects, they seem a bit distant and dry, always theorists looking at the big picture, while Archimedes is a much more comfortable role model for the modern experimentalist.

Conclusion

By the time the Greek world came under the control of Rome, a powerful group of Greek thinkers had completed the creation of the study of nature as a discipline and removed all but the most tangential connection to supernatural beings or forces. They made the universe measurable, and thus it could be known. They set the framework for intellectual inquiry that would be used in the Mediterranean world for over 1,000 years, and a number of ideas from Aristotle and Plato still provoke debate to this day. Under Roman control, Alexandria became even more important as a centre of learning, and the basis of Aristotelian philosophy was exported to the far flung reaches of the Empire, from Roman Britain to the Fertile Crescent in the Middle East. Along with the philosophy went a new image of the sage, the scholar, the intellectual, whose job was not to interpret the mysteries of a world full of spirits but to read and reveal the text of the book of nature.

Essay Questions

1.   Why did natural philosophy develop in the Greek world rather than in Egypt or the Fertile Crescent?

2.   What were the principle concerns of Greek natural philosophers?

3.   Comparing Plato's and Aristotle's systems, what were similar concerns and how did they differ?

4.   What was Aristotelian logic and why was it so important for natural philosophy?

CHAPTER TIMELINE

753 BCE	Traditional date for founding of Rome
146 BCE	Greece comes under Roman control
C. 50 BCE	Lucretius writes *De rerum natura*
31 BCE	Egypt comes under Roman control
C. 7 BCE	Strabo writes *De situ orbis*
79 CE	Pliny the Elder writes *Natural History*
C. 148	Ptolemy writes *Almagest* and *Geographia*
162	Galen moves to Rome
250	
286	Roman Empire divided into east and west
392	Christianization of Roman Empire
476	End of Roman rule in western empire
529	Lyceum and Academy closed by Emperor Justinian
622	Muhammad travels to Yathrib (Medina). The *Hegira* marks beginning of Islam
762	Baghdad established
C. 815	*Bait al-hikmah* (House of Wisdom) created
C. 820	al-Kindi translates *Geographia*
C. 890	al-Razi writes *Secret of Secrets* alchemical text
C. 900	Founding of Balkhi school of geographic thought
950	
C. 1000	Almost all surviving Greek medical and natural philosophical texts translated into Arabic
C. 1037	Ibn Sina dies
1426	
1520	

Mayan Empire flourishes — (bracket spanning 250–950)

Aztec Empire flourishes — (bracket spanning 1426–1520)

THE ROMAN ERA AND
THE RISE OF ISLAM

W hile the Greek philosophers were struggling with the structure of the cosmos, across the Adriatic Sea a small group of people living on the east bank of the Tiber River were in the process of creating a powerful military state. Traditional legends claim Romulus and Remus founded Rome in 753 BCE, but the origins of the city were probably Etruscan. Around 500 BCE Etruscan rule ended and Roman rule began. Rome expanded its area of control through the fourth and third centuries BCE, conquering or absorbing its neighbours. When Rome fought the Punic Wars against Carthage between 264 and 146 BCE, it established its military prowess and began its rise to empire.

As Rome expanded, it came into contact with Greek culture both through Greek colonies on the Italian peninsula and later by conquest of Greece itself. Roman dominance of Greece was completed by 146 BCE, and with the occupation the intellectual heritage of Greece came largely under the control of the Roman Empire. Greek scholarship was not destroyed by Rome, and in fact the Roman elite adopted Greek education and studied Greek philosophy, holding many Greek philosophers in high regard. This regard was not generally for the sake of philosophy but for a more practical purpose. Mastering Greek philosophy was seen as a good method to discipline the mind just as the legionnaires disciplined the body; both prepared the elite of Rome for their role as masters of the world. The Romans were at heart a people interested in practical knowledge. Their engineers created buildings, roads, aqueducts, and many other magnificent structures that have

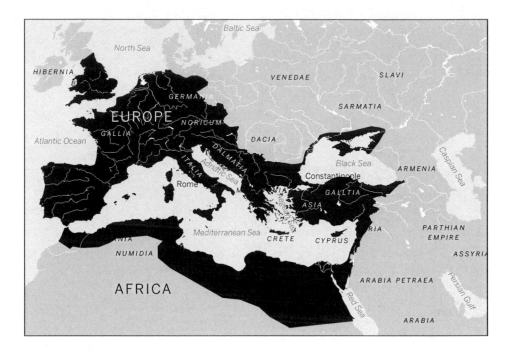

2.1 THE ROMAN EMPIRE

survived into the modern world. As impressive as the end products of Roman industry were, even more important was the power of the organizational system that could conceive, manage, and expand the enormous empire. In the Roman Empire nature was to be bent to useful ends.

The study of nature for the Romans was, therefore, oriented more toward practicality than philosophical speculation. Roman intellectuals were more concerned that a thing worked than about demonstrating the truth of the knowledge of that thing. Thus, they were more concerned with machines, studies of plants and animals, medicine, and astronomy than epistemology or philosophy. The Roman Empire was not based, as the Greek city-states had been, on public discourse and democracy but on public demonstrations of power. Making nature do your bidding was more essential than right reasoning. The Romans took the Greek heritage, in natural philosophy as in much else, and transformed it to aid their own objectives.

For the Roman elite, learning Greek philosophy might not be an end in itself but a way of training the mind. Intellectual acuity, even if the ends were material, still required a sound foundation. This heritage led a number of Roman intellectuals to preserve and propound Greek thought. For example, around 75 BCE the famous orator and politician Marcus Tullius Cicero (106–43 BCE) located and restored the tomb of Archimedes, who despite fighting against the Romans was well liked for his facility with machines. In 50 BCE the poet Titus Lucretius Carus (c. 95–55 BCE) wrote *De rerum natura*, a defence of Epicurean philosophy, and expounded the theory of Democritean atomism. In 40 BCE Marc Antony gave some 200,000 scrolls (primarily from the library at Pergamum) to Cleopatra (a descendent of the Greek rulers established by Alexander), who added them to the library of the Museum at Alexandria, making it the largest collection in the world. The gift was not completely altruistic, as Antony hoped to extend Rome's influence in Egypt, but it certainly confirmed the library's value. When Rome subjugated Egypt in 31 BCE, the conquerors, well aware of the Museum's worth, preserved the greatest centre of learning in the Mediterranean world both as an ornament in their empire and for the practical value of its materials.

The Romans developed a taste for large-scale projects. One of the keys to their success was the widespread use in their architecture of the arch, which allowed them to create much larger and much more open structures than the Egyptians or Greeks had been able to build using the column and lintel system. An arch rotated in three dimensions produces a dome, which was another innovation in Roman architecture. They also introduced the use of hydraulic cement as a mortar; because it set even under water, it was a very useful tool for building bridges, piers, and docks.

The greatest engineering accomplishment of the Roman era was the road system. While the majority of Roman roads did not represent the most complex engineering problems that had to be mastered, they were the key to the centralized control of the empire. Roman power functioned because the roads not only provided a communications system and a safe trade route but also allowed the rapid deployment of military forces.

Natural Philosophy in the Roman Era: Ptolemy and Galen

While Roman engineers were inventing and developing solutions to the problems of empire, natural philosophers in the Roman era were not as innovative. They did not create a new system of natural philosophy but turned their energy to continuing and extending the philosophical systems that came from Greece, particularly the Aristotelian (which dominated at Alexandria) and the Platonic systems. One way this extension took shape was in the commentaries and encyclopedic work of a number of scholars such as Posidonius (c. 135–51 BCE), who wrote commentaries on Plato and Aristotle, and Pliny the Elder (23–79 CE), whose massive work *Natural History* was nothing less than a complete survey of all that was known about the natural world presented to an educated but general audience. It was reported that Pliny and his assistants reviewed more than 2,000 volumes to compile their information. Some of this material was fantastic and mythical, such as descriptions of strange beasts and people with no heads, but Pliny also reiterated Eratosthenes' measurement of the size of the globe.

Two exceptions to the largely derivative natural philosophy in the Roman era were advances in astronomy and medicine. In both cases, the philosophical foundation came primarily from Aristotle but was extended well beyond any work of the earlier Greek period. In addition to the importance of the work itself, both the astronomy of Ptolemy (c. 87–c. 150 CE) and the medical discoveries of Galen (129–c. 210 CE) were significant conduits for the transmission of Greek philosophy to scholars after the fall of Rome.

Ptolemy's Astronomy

Although we know almost nothing about Ptolemy's life, his work is recognized as the cornerstone of natural philosophy to this day. His full name was Claudius Ptolemaeus, which suggests both Greek and Roman roots. Living in Alexandria, he produced material on astrology, astronomy, and geography, using complex mathematics and a large body of observations. His methods of astronomical calculation in particular shaped the Western view of the heavens for more than 1,000 years. In terms of accuracy, his observations were not surpassed until the beginning of the seventeenth century in the era of Tycho Brahe and with Galileo's introduction of the telescope.

Ptolemy's work on astronomy, collected in the *Mathematical Syntaxis*, commonly known as the *Almagest* (from the Arabic *al-majisti* meaning "the best"), accomplished two things. First, he created a mathematical model that reconciled Aristotelian cosmology with observation. Second, he provided a comprehensive tool, including tables and directions, to make accurate observations. His work extended both that of Hipparchus of Rhodes (fl. second century BCE), who had made numerous precise observations of the stars and planets, worked out the precession of the equinoxes, and measured the length of the year and the lunar month; and of Eudoxus (c. 390–c. 337 BCE), whose system of nested spheres each with a slightly different axis of rotation was a creative solution to the problem of retrograde motion.

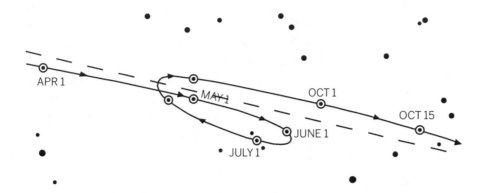

2.2 RETROGRADE MOTION

Aristotle's geocentric or Earth-centred system seemed to be obvious from general experience and is philosophically consistent, but there were several problems with it when it came to detailed observation. One of the most difficult observations to reconcile was retrograde motion. (See figure 2.2.) If an observer traced the course of the planets Venus, Mercury, Mars, Jupiter, and Saturn over an extended time relative to the stars (which move eastward in a yearly cycle), each planet gradually moved eastward, and then seemed to slow down and loop back westward for a time before continuing their west to east movement. This was most noticeable in the orbit of Mars.

In addition to the problem of retrograde motion, a number of the planets seemed to move at different speeds in different parts of their orbits, while the fixed stars moved in a very regular pattern. The combined problems of motion and time seemed to contradict the axiom of the perfect circular and spherical nature of the heavens. Retrograde motion also presented practical problems, since precise knowledge of the objects of the skies was necessary for casting horoscopes, aiding navigation, and telling time. Ptolemy created a working model of the heavens that resolved all these problems. It is important to understand that he regarded his model not as a true description of the universe but rather as a mathematical device that allowed observers to track the movement of the celestial bodies. Because of the utility of his system and its fit with the philosophy and theology of later scholars, the Ptolemaic system became synonymous with the actual structure of the heavens.

Ptolemy based his deductions on a large body of observations that came from the resources of the Museum's library and from his own work and that of assistants. To reconcile the necessity of circular motion (as required by Aristotelian cosmology) with the observed motion of the planets, he introduced geometric

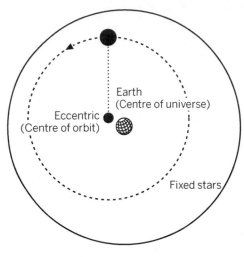

2.3 ECCENTRIC

"fixes" that allowed for a mechanization of the process of mapping the movement of the celestial bodies. These fixes were the eccentric, the epicycle, and the equant.

If a planet moved uniformly around the Earth, there was no problem describing its orbit as circular, but most planets seemed to move differently at different points in their orbit. Ptolemy reasoned that the problem could not be the planet actually going faster and slower (what mechanism could cause such a change?), but our perception of the motion. By moving the centre of the planet's orbit away from the Earth (at the centre of the universe), the eccentric replicated the observed non-uniform motion while allowing the planet to follow a perfect circular orbit in uniform motion. (See figure 2.3.)

The eccentric did not solve all the problems, so Ptolemy also introduced the epicycle, a small circle centred on a larger circle or deferent. (See figure 2.4.) This fix neatly accounted for retrograde motion. Later astronomers realized that epicycles could be added to solve observational problems, as we see particularly in late medieval and Renaissance mechanical models.

The equant was the most complex of Ptolemy's devices. (See figure 2.5.) The equant is not at the centre of the orbit but is displaced from it. However, the motion of the planet on the deferent is uniform around the equant. This means that the planet's apparent motion will be faster and slower in different parts of the orbit because the region swept out by the planet will not be equal.

Using these three geometric devices, Ptolemy was able to account for all the varied motions of the heavens and to predict future celestial activities. The *Almagest* was a brilliant achievement, and his system was so powerful that it became the basis for Western and Middle Eastern astronomy for over 1,300 years; a version of it survives to this day for small craft navigation at sea. Although much of the *Almagest* was complicated, part of its power was that it was not mathematically complex. All the elements of Ptolemy's models were based on the geometry of the circle, which was well understood. While there could be many epicycles employed to establish the orbit of a planet, they were all constructed the same way. The *Almagest* provided a complete account of celestial motion of all the objects that could be seen with the naked eye. The observations were so accurate and the method of calculation so complete that from a practical point of view Ptolemy had resolved the issue of astronomy. There was some tinkering with the distribution of epicycles and the exact location of the eccentrics, but the model worked so well that it could be made into a mechanical device. This celestial clock was perfected by Giovanni de Dondi of Padua around 1350 CE, and a working copy of his masterpiece of clockwork and Ptolemaic astronomy can be found at the Smithsonian Institution in Washington, DC.

Ptolemy's other great work, the *Geographia*, applied his powerful mathematical tools and the resources of the Museum to the terrestrial realm. In a sense, the *Almagest* and the *Geographia* represent two parts of the same system, the first representing the supralunar realm and the second the terrestrial or sublunar realm. To achieve good astronomical results it was necessary to know where you were on the globe; to know that, the globe had to be treated mathematically. Ptolemy summarized the work of other geographers and examined aspects of cartography including various methods of projection, longitude, and latitude; he then provided lists of some 8,000 places and their coordinates. He treated the celestial and terrestrial globes as equivalent, applying the same grid system to each, and using the same spherical geometry to plot points. He divided the globe into a series of parallel belts or "climates" and developed a grid of longitude and latitude coordinates. In doing so, he created a map projection that has never been completely superseded and that was of immense importance to later European exploration and contact with other parts of the world.

Ptolemy's mathematical geography contrasts with the earlier descriptive geography of Greek scholars such as Strabo (c. 63 BCE–c. 21 CE). Strabo wrote an eight-volume geography, *De situ orbis*, around 7 BCE, in which he set out to describe every detail of the known world, based both on his own extensive travels and on the accounts he gathered from other travellers. This was an enterprise closely tied to history and politics. Ptolemy made a distinction between his mathematical rendering of the globe, which he called "geography," and Strabo's type of terrestrial research, labelled "chorography."

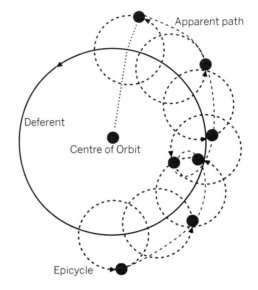

2.4 EPICYCLE

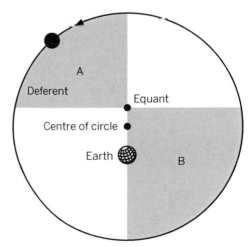

2.5 EQUANT
The orbiting planet takes equal time to travel around quadrant "A" as around quadrant "B."

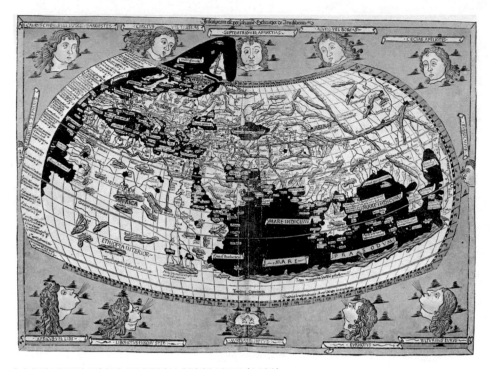

2.6 PTOLEMY'S WORLD MAP FROM *GEOGRAPHIA* (1482)

The most useful parts of Ptolemy's *Almagest*, as well as his "Table of Important Cities" from the *Geographia*, were compiled as the *Handy Tables*. This reference allowed a quick way of doing celestial calculations and was easier than the methods outlined in the *Almagest*. The *Handy Tables* became a standard tool for astronomy. The geographical material was not as well known, nor as widely circulated, as the astronomical, and it faded from sight after the fall of Rome. Its rediscovery in the fifteenth century had a major impact on geographical thought and exploration in Renaissance Europe. Because Ptolemy's works were so useful, they were widely disseminated, which in turn helped them to survive the turmoil of the end of a number of empires including the Roman and the Byzantine. Where Ptolemy's work survived, the Aristotelian foundation of his work also persisted.

Galen's Medicine

In contrast to Ptolemy, we know much more about Galen's life. Born in 129 CE at Pergamum, second only to Alexandria as a centre of learning in the period, Galen studied mathematics and philosophy before beginning his medical training at the age of 16. In 157 CE he became surgeon to the gladiators at Pergamum. In many ways, it was this first professional work that allowed him to begin creating his own system of medical knowledge, particularly of anatomy. At a time when human dissection was forbidden, he got first-hand experience of human anatomy by tending to wounded and dead gladiators. He saw the structure of muscle and bone, sinew and intestine laid bare by violent injury and was responsible for trying to set the parts back in place when possible. In 162 CE he travelled to Rome, remaining for four years before returning to his home town. When a plague struck, Emperor Marcus Aurelius called him back to Rome, where he settled permanently as the personal physician to four emperors: Marcus Aurelius, Lucius Verus, Commodus, and Septimius Severus.

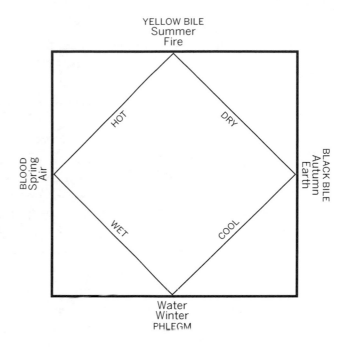

2.7 THE FOUR GALENIC HUMOURS

Medical philosophy of the time was dominated by Hippocratic theory. Hippocrates of Cos (c. 460–c. 370 BCE) may have been a single individual, a mythic figure, or a name given to a collective. The Hippocratic system of medicine was based on the concept of regimen and balance. Regimen covered not only the physical aspects of a patient's health but also social, mental, and spiritual aspects. Hippocratic doctors conducted long interviews with patients, asking about their diet, work, home life, sex life, and spiritual health. Horoscopes were cast, and even geography was considered, since living near a swamp or exposure to certain winds was considered harmful to well-being. Although Hippocratic doctors considered spiritual aspects and used horoscopes, they regarded illness as primarily natural rather than supernatural and thus treated disease with material solutions such as drugs, diet, and exercise. Health was considered to be the correct balance of physical action, diet, and lifestyle; illness represented an unbalancing of the elements.

The basis of the Hippocratic theory of balance was the four humours, which covered the four bodily fluids: phlegm from the head, blood from the liver, yellow bile from the stomach, and black bile from the stomach or intestine. (See figure 2.7.) Each of these had a paired quality of hot/cold and wet/dry as well as one of the four elements that fit nicely with the Aristotelian system. The objective of medical intervention was to balance the four humours. Too much or too little of any humour resulted in illness. For example, a person with too much blood was sanguine and the treatment was bleeding, while a bilious person had too much bile in their system and needed a purgative. These physical conditions were, as the modern use of the terms sanguine and bilious suggest, also associated with temperament.

While modern medicine often traces its heritage back to the Hippocratic doctors of Greece, especially through the Hippocratic Oath, there were in fact few unified theories of medicine in Greek or Roman times. Although some Hippocratic doctors had experience with medical trauma such as wounds, fractures, and other injuries, there were other medical practitioners, such as surgeons, who dealt with the body directly. Hippocratic doctors did not deal with women, who were treated by another group of practitioners including midwives. To a large extent, medical treatment for men and women was separate. If a man could afford a physician, he sought a philosophically trained physician like Galen. For women and the poor, there were a range of practitioners and a range of treatments from the most practical to the most spiritual. Aid was often provided to the sick at temples, where prayer and supplication were part of treatment.

At least four general groups of medical philosophy have been identified by historians at the time Galen began to practise medicine: the rationalist, empiricist, methodist, and pneumatist. Even within those groups there was no united form of practice. Each school and each doctor who took on apprentices taught a different version of medical theory. Further, each doctor was also his own salesman seeking clients and patrons, often literally in the marketplace. Because of this competition for clients, a doctor had to be able to persuade potential clients that his brand of medicine was the best, so medical education also included training in philosophy, rhetoric, and disputation.

When Galen became physician to the gladiators of Pergamum, he was taking a lucrative job but one with relatively low status. Gladiatorial combat was a big money enterprise, but Galen's work, primarily dealing with wounds, was considered very practical and thus of a lower status than the intellectual diagnosis of disease. Despite the issue of status, treating the gladiators gave Galen something he could not get elsewhere—detailed exposure to human anatomy. Religious and cultural taboos prevented the dissection of humans, so in the Roman world anatomical training was theoretical or based on animal dissection, particularly of monkeys.

Galen brought to his work his philosophical training, which covered Plato, Aristotle, and the Stoics, who believed in a physics based on the material world, as well as many other elements of classical thought. He accepted the Hippocratic humours but wanted to make clear the functions of the human organs, so he applied Aristotelian categories, particularly the four causes, to his anatomical work. His close observation demonstrated, for example, that arteries carried blood rather than the older theory that they carried "pnuema," or air. Each organ and structure in the body had a purpose, and dissection and vivisection were the key tools to establishing what that purpose was. His anatomical work became a powerful tool not only for physiology but also for persuasion. His demonstrations put him ahead of other physicians who depended on rhetoric to sell their brand of medicine, because he could show people his system through actual dissections of animals. Galen's success was so great that he received the patronage of four emperors and was physician to the elite of Roman society. The support afforded by such patronage allowed him the time to write, and he may have authored as many as 500 treatises, of which more than 80 survive to the present.

Galen's productivity and patrons would have been enough to ensure him a place in medical history, but a further element helped preserve Galenic medicine when other aspects of Greco-Roman thought were repudiated or lost. Galen adopted a strongly teleological philosophy in which nothing existed without a purpose and all of nature was constructed in the best possible way and for the best possible good. This demonstrated the existence and perfection of the Demiurge, the fashioner of the world. While extending Platonic idealism to the body, this teleological philosophy fit well with that of Islamic, Jewish, and Christian thinkers. His practical medicine was considered one of the few things worth keeping from the pagan, decadent, and materialistic world of the Roman Empire by these religions, which had strong precepts about care for the sick as a religious duty. When Galenic texts survived, the philosophical foundations of Plato and Aristotle were also preserved.

The Decline of Rome

By the time of Galen's death around 210 CE, cracks were beginning to appear in the Roman Empire. In 269 CE, the library of Alexandria was partly burned when Septima Zenobia captured Egypt. In 286 CE, Diocletian divided the empire into eastern and western administrative units. Alexandria's great library was burned to the ground in 389 during a riot between pagans and Christians, which closed the Museum. Emperor Theodosius I ordered the destruction of pagan temples in 392 CE as the last vestiges of Roman religion were officially replaced by Christianity.

The Roman Empire might have continued to weather these problems had it not been for pressure from the outside. In the east the Persians resisted Roman control and were a constant threat. In the northwest a whole host of peoples were either pushing against the frontier at the Rhine or struggling against Roman control. Attacks across the frontier increased in the fifth century because the Germanic tribes were being pushed from behind by the Huns while at the same time being attracted to the wealth of the faltering empire. The Roman roads, which had once been the means of controlling the empire, guided the invaders into the heart of the Roman world.

In the eastern part of the empire Constantinople struggled to continue the traditions of empire and learning after Rome itself had fallen to barbarian invaders. Established in 330 CE by Emperor Constantine I, it became the capital of a separate eastern empire in 395 CE after the death of Theodosius I. With its Greek heritage and strategic location, Constantinople preserved elements of Greek learning and culture during the years of decline caused by the end of central rule by Rome.

Rome, once the greatest city of an empire that spanned much of Europe, northern Africa, and the Middle East, was sacked by the Visigoths in 410 CE, then again by the Vandals in 455 CE. The fall was partly the result of internal problems such as increased levels of taxation, the disparity of rich and poor, the grievances of the conquered peoples, a decline in Roman participation in the military, and constant political infighting and civil war. Romans' ability to deal with these problems may have been significantly diminished by the effects of lead poisoning, as the heavy metal leached not only from alloy plates, bowls, and cups but also from the lead pipes extensively used for plumbing (*plumbum* being the Latin for lead). Moreover, acetate of lead was used to sweeten wine.

The western empire finally ceased to function in 476 CE when German invaders established themselves as the emperors of Rome. The new rulers were not particularly interested in philosophy, natural or otherwise. The last great natural philosopher in Rome was Anicius Manlius Severinus, better known as Boethius (480–524 CE). His best remembered work was *On the Consolation of Philosophy*, but his most lasting impact came from his role as a translator. He translated a number of Aristotle's logical works into Latin, as well as Porphyry's *Introduction to Aristotle's Logic* and Euclid's *Elements*. Because of his work, Aristotle was almost the only available source for Greek natural philosophy until the twelfth century. Boethius was jailed for treason and finally executed by Theoderic the Great in 524 CE, ending Greek-oriented natural philosophy in Western Europe for almost seven centuries.

Early Christianity and Natural Philosophy

The other element that resulted in a decline in interest in Greek philosophy was the rise of Christianity. Whether it also contributed to the decline of the Roman Empire is a matter of historical debate. On the one hand, Christians were frequently associated with dissident elements in Roman society, and efforts by the government to fight the spread of the religion at home took resources and attention away from other problems such as the barbarian pressures on the borders. On the other hand, the Christians did not create the external problems, and with the Christianization of the Roman Empire under Theodosius I starting in 392 CE, Christianity offered the possibility of a unifying force in a badly fractured society. In the short term, the Roman Christians suffered during the collapse of the western empire, but in the longer term the Church was able to preserve and rekindle the intellectual aspects of philosophy.

Christianity had (and continues to have) an uneasy relationship with natural philosophy. The messianic and evangelical aspects of the religion pointed people away from the study of nature and toward the contemplation of God. The Greek philosophers were pagans and, therefore, to be rejected, but they were also part of the extraordinary Roman Empire and closely linked with the intellectual power and managerial skills, particularly literacy and bookkeeping, that the Church needed to survive. In addition, many of the most important leaders of the early Church were trained in Greek philosophy. Augustine (354–430 CE) in particular was well trained in Greek philosophy, and he became the voice of the intellectual Church. Even so, the last vestiges of Greek philosophical education, the Academy and the Lyceum, were closed by Emperor Justinian in 529 CE. Although they had been little more than tattered remnants of their former glory for years, the closing marked the end of Greco-Roman philosophical power.

Christianity was fraught with internal controversy, and the problem of heresy was frequent. The Donatists' and Arians' challenge to the theology that emanated from Rome as early as the third century led to the Council of Nicaea in 325 CE and the promulgation of the Niceaen Creed to establish orthodoxy. Even ownership of

Bibles became an issue, and in many places only the clergy were legally allowed to own them. With literacy largely confined to members of the Church, this was not a difficult restriction to enforce, and the lack of literacy meant that Greek material was equally inaccessible.

Through deliberate and accidental destruction, loss, or rejection, the intellectual heritage of natural philosophy largely disappeared in the Latin West (the western lands of the Roman Empire where Latin was the language of the Church and the small educated class). Some texts and ideas survived in scattered pockets in the West, while in the eastern empire (where Greek was still used) Byzantium held on to more material. What survived were parts of the work of Hippocrates and Galen on medicine, because of the duty of the Church to care for the sick; parts of Euclid; fragments of Aristotle's logic; parts of Plato, particularly *Timaeus*; and some of the ideas of astronomy from Ptolemy, which were used to help keep up the calendars. Many Christians thought they were living in the end days of biblical prophecy, so there was little impetus to preserve or study the old knowledge even if it were available. While the light of philosophy never went out completely in the West, it was dimmed considerably, and what remained was folded back into theology.

The Rise of Islam and Its Effect on the Development of Natural Philosophy

The void created by the collapse of Rome also had an effect on the southeast side of the Mediterranean. The people who lived on the Arabian Peninsula and in the Middle East were conveniently placed to trade with Asia, Africa, and Europe, and a number of important centres developed that had access to both Persian and Byzantine markets. Territorial conflicts as well as struggles based on cultural and religious differences resulted in the emergence of a number of independent states in the region. In this period of turmoil, Muhammad began his efforts to convert the people of Arabia to his new religion. Unable to gain significant inroads in Mecca, he travelled to Yathrib (later Medina) in 622. This trip, called the *Hegira*, marked the first step toward the foundation of the Islamic world.

Muhammad's return to Mecca in 630 was followed by a wave of conversions both through peaceful exhortations by preachers and traders and by the sword and the *jihad* or holy war. By the time of his death in 632, most of the Arabian Peninsula had been converted to Islam. Syria (previously a Byzantine province),

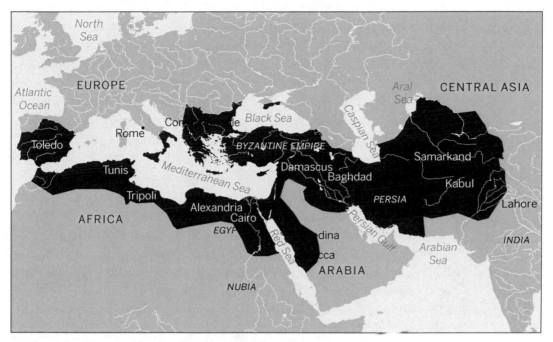

2.8 ISLAMIC AND BYZANTINE EMPIRES 750–1000

Palestine, and Persia followed by 641; a year later, Egypt came under the control of the first caliphs, who were both political and religious rulers. Under the Umayyad caliphs the conquests continued, and by 750 the Islamic empire ran from Spain in the west to the Indus River in the east.

The capture of many of the most important centres of learning in the Middle East, particularly Alexandria, gave Islamic scholars control of vital intellectual resources. As the strength of the Arabic world grew, so did Islamic scholarship, first translating and integrating the philosophies of the Greco-Roman world and then establishing a very high level of competence in research and critical analysis. Yet, until recently, Western historians of science have regarded Islamic natural philosophers as little more than imitators of Greek work and a conduit through which it passed to European scholars. So clear was this prejudice that many older history texts used only the Latin version of Arabic names, thereby suggesting that the only significant Islamic works were those used by Western European scholars. The reason for this dismissal or denial of authentic and innovative study was the idea that Islamic natural philosophers, despite access to the Greek material, failed to expand upon it, whereas the same material in the hands of European scholars led to a revolution in natural philosophy and the creation of modern science.

More recent scholarship awards Islamic thinkers a far greater role in shaping the work of later natural philosophers. Islamic scholars did not accept Greek thought unchallenged and added not only their critical thinking to the body of material available but also their own original research. They were also far more willing to test ideas than the Aristotelian or Platonic philosophers had been, and although this should not be confused with experimentalism (which uses a different philosophic conception of certainty), it became an acceptable tool for natural philosophy because of its use by these scholars.

As in Christianity and Judaism, there was a tension between the intellectual and spiritual aspects of Islamic theology, but certain of its tenets made it amenable to the study of nature, particularly if the elements of faith were interpreted broadly. One of the five pillars of the faith was *Shahadah*, the profession of the creed, which essentially called all the faithful to read the Q'ran, the holy book of Islam. This resulted in a push toward literacy and the promotion of Arabic as a unifying language from the Iberian Peninsula in the west to the border with China in the east. A second pillar, the *Hajj*, or pilgrimage to Mecca, brought together the people of the far-flung empire, even when different regions were not politically unified. This created ties of personal contact, trade, and intellectual exchange. Indian mathematics, Chinese astronomy and inventions, and Hellenized Persian culture flowed up and down the pilgrimage and trade routes along with silk, ivory, and spices. The most famous of these arteries was the Silk Road, the lengthy trade route that connected China to the Arabic world. While the Silk Road is best remembered for the exotic products that moved from east to west, it also brought ideas including Hindu mathematics and Chinese alchemy.

The Q'ran itself also contributed to a more positive attitude toward the study of nature than the Bible did for Europeans. It was more precise about creed and liturgy (reducing the potential for schism), but was also more worldly, calling on the faithful to study nature as part of God's creation. Many of its passages present knowledge and the acquisition of knowledge as sacred. One of Muhammad's most famous sayings was "Seek knowledge from the cradle to the grave." The centre of Islamic religious life was the mosque, which, particularly outside the Arabic regions, served as a school of Arabic literacy. Many mosque schools, or *maktab*, developed into more extensive educational institutions and became essentially the first universities, offering advanced studies for students and research facilities, such as libraries, for scholars.

Another aspect that should not be overlooked was the sheer wealth of the Islamic empires. The ability of caliphs to order the creation of schools, libraries, hospitals, and even whole cities demonstrates their economic power. With those resources available, even a low level of interest in natural philosophy could produce significant results. The Islamic world received large collections of Greek and Roman material along with their conquests, and its proximity to the Byzantine Empire meant, at least in times of peace, a potential for intellectual exchange. Educated Persians and Syrians, with their knowledge of Greek culture running back to the time of Alexander the Great, became bureaucrats within the empires and brought with them their intellectual heritage.

The Islamic Renaissance

When a new dynasty started under the Abbasids, there was increased interest in the intellectual heritage of the Greeks. The early Abbasids were intellectually tolerant and had a strong interest in practical skills, employing educated Persians and even Christians in government. In particular, the Nestorians (a Christian sect from Persia) served as court physicians. They practised Galenic medicine, preserving not only the practical aspects of Galen's work but its Aristotelian and Platonic foundation as well.

In 762 the Abbasid Caliph al-Mansur established a new capital, Baghdad, on the Tigris River. He also began a tradition of translation of Greek and Syriac texts into Arabic. His grandson, Harun ar-Rashid, continued this work and even sent people to Byzantium to look for manuscripts. However, the greatest intellectual developments came under Harun's son, al-Mamun, who around 815 created the *Bait al-hikmah* or House of Wisdom. This was part research centre, containing an extensive library and an observatory, and part school, attracting many of the most important scholars of the day. This state-supported enterprise was also responsible for the majority of the translation of Greek, Persian, and Indian material into Arabic.

The head of al-Mamun's research centre was Hunayn ibn Ishaq (808–873), a Nestorian Christian and physician, who grew up bilingual (Arabic and Syriac) and later learned Greek, perhaps in Alexandria. He translated over 100 works, many of them medical. His son and other relatives continued the translation work, in particular Euclid's *Elements* and Ptolemy's *Almagest*, both of which became important foundational texts for Islamic scholars. By 1000 almost every surviving work of Greek medicine, natural philosophy, logic, and mathematics had been translated into Arabic.

The interest in education in Islam fostered the appearance and high status of the *hakim*, a sage or wise man, and philosophers such as Aristotle were revered as sages. The educational system included philosophy and natural philosophy as components of a well-rounded education. There was a great flowering of culture, known as the Islamic Renaissance, starting in the ninth century and running until about the twelfth century. During this period Islamic scholars continued the intellectual traditions of the Greeks, but there were important differences. Islamic scholars had to conduct their work within the framework of their religion. While there were liberal and conservative periods, often varying with a change in rulers, Greek material could not simply be adopted outright. Some aspects were accepted with little change, such as Ptolemaic astronomy; some were modified, such as the introduction of God rather than an indefinite "unmoved mover" in Aristotelian physics; and some elements were rejected outright, such as various cosmological creation stories that came from Greek and Roman sources.

In addition to the questioning inherent in the ratification of pagan material, Islamic scholars pursued new ideas in natural philosophy. This was partly a result of circumstances, since scholars often lacked access to the complete corpus of Greek thought and so might have only a fragmentary idea of, for example, Aristotelian optics. It was then necessary to do independent work on the topic. Islamic scholars were also more interested in testing observations than Aristotle or Plato had been, in part because they had a less intellectualized concept of the acquisition of natural philosophic knowledge. In other words, they had a more hands-on approach. This attitude toward knowledge acquisition coincided with expectations for the educated class in Islamic society, since the educated and affluent were supposed to be able to turn their hand to poetry and music, history and philosophy, and martial arts such as riding and swordplay, as well as understanding practical matters such as commerce and trade. Scholarship and courtly behaviour were intimately linked in the lives of many of Islam's greatest natural philosophers, and many of the traits associated with chivalry for the European knights were in fact adopted from the Islamic world.

Another reason Islamic scholars were more willing to test nature was because they lived in a more materially oriented and technically advanced society. The craft skills of the Arabic world were extremely accomplished, surpassed in this period only by China, which was a trade partner. Artisans made a wide range of tools and instruments, and there was both an appreciation for fine work and the money to support it. Two examples of this high level of skill can be seen in glass-making and metallurgy. Glass-making was a large-scale industry that produced many of the tools used by Islamic scholars to investigate optics and alchemy, while metalworkers produced instruments such as astrolabes and armillary spheres. Another development in metalwork that intrigued (and terrified) Europeans was Damascus steel, which in the form of swords gave the armies of Islam an edge (literally and figuratively) over the weapons used by the Crusaders.

Many of the greatest Islamic natural philosophers were educated as physicians, which perfectly combined practical and theoretical training. This meant that they were first introduced to Greek philosophy through Galenic material. While a distinction existed between the intellectual understanding of health and disease and the practical matters of surgery and bone-setting, the technical skills of Islamic practitioners surpassed those of the Greco-Roman world and far outstripped their European neighbours. Technical abilities and tools extended to abdominal surgery and cataract removal. Eye surgery was linked to theories of vision and the more theoretical study of optics. Thus, medicine was a perfect conduit for natural philosophy in the Islamic world. It was theologically sound, since care for the ill was part of the charity requirements of the faith, and it was both practical and intellectual without being a craft, and thus acceptable for the upper class. With these characteristics, physicians frequently held high posts in government and at court.

Agriculture was another area of expertise for Islamic scholars and practitioners. The coming of Islam freed many farmers from their previous overlords; this freedom combined with increased literacy encouraged a burgeoning of practical and theoretical work on agriculture and botany. In part because of the lines of communication that were established within the Islamic world, and in part because of the freedom that farmers enjoyed (in comparison to the peasants of Latin Europe), interest in useful plants led to one of the largest transfers of biological material in history, as crop plants and their particular farming needs were transferred from China in the east throughout the Islamic world to the Iberian Peninsula in the west. A partial list of transplanted crops includes bananas, cotton, coconut palms, hard wheat, citrus fruit, plantain, rice, sorghum, watermelons, and sugar cane. A somewhat less practical biological exchange occurred in 801 when Caliph Harun al-Rushid of Baghdad sent an elephant as a present to Charlemagne. The collecting of plants, both useful and decorative (roses, tulips, and irises were also part of the great plant transfer) led Islamic scholars to create encyclopedias of plants such as al-Dinawari's (828–896) *The Book of Plants* and Ibn al-Baitar's (c. 1188–1248) *Kitab al-jami' li-mufradat al-adwiya wa al-aghdhiya*, a pharmacopoeia listing over 1,400 plants and their medicinal uses. One of the world's largest botanical gardens was established in Toledo in the eleventh century.

One of the most powerful minds of the Islamic Renaissance was Abu 'Ali al-Husain ibn Abdallah ibn Sina (980–1037), whose life was chronicled in his autobiography and the memoirs of his students. He was a child prodigy who had memorized the Q'ran by the age of ten and had begun training as a physician when he was 13, although he also studied widely in philosophy. After curing the Samanid ruler Nuh ibn Mansur of an illness, he was allowed to use the Royal Library. It was then that ibn Sina began to explore a vast range of material from mathematics to poetics. Because of his skills as a physician, he found employment at the courts of various rulers, but it was a turbulent time, and he was involved in a number of political struggles that saw him made a vizier by Prince Shams ad-Dawlah in Hamadan, a region in west-central Iran, only to be forced from office and jailed for a time.

Ibn Sina left Hamadan in 1022, on the death of the Buyid prince he had been serving, and moved to Isfahan. He entered the court of the local prince and spent the last years of his life in relative calm, completing the major works he began in Hamadan. He was prolific, producing over 250 works covering medicine, physics, geology, mathematics, theology, and philosophy. He wrote so much that he had a special pannier made so he could write while on horseback. His two most famous books were the *Kitab al-Shifa'* (*The Book of Healing*) and *Al Qanun fi al-Tibb* (*The Canon of Medicine*). Despite its title, *Kitab al-Shifa'* is actually a scientific encyclopedia covering logic, natural philosophy, psychology, geometry, astronomy, arithmetic, and music. Although including many aspects of Greek thought, particularly Aristotle and Euclid, it does not simply recount those works. The *Al Qanun fi al-Tibb* became one of the most important sources of medical knowledge. It was both a translation of and a commentary on Galenic medicine and contains what is perhaps the first discussion of mental illness as a form of disease. When ibn Sina's work was discovered by Latin scholars, his name was translated as Avicenna, and his books helped fuel a drive to rediscover Aristotle.

A contemporary of ibn Sina was Abu Ali al-Hasan ibn al-Haytham (c. 965– c. 1039). Although ibn al-Haytham was not trained as a physician, he worked on vision, diseases of vision, and the theory of optics. In his *Kitab al-Manazir* (*Book of Optics*) he presented the first detailed descriptions and illustrations of the parts of the eye in optical terms and challenged the Aristotelian optics of Ptolemy. Where Ptolemy had supported the extramission theory of vision that was based on a kind of ray coming out of the eye and intersecting objects to produce sight, ibn al-Haytham supported the intromission theory that posited that light struck objects and that rays then travelled from the object into the eye. He also described refraction mathematically and undertook a series of experiments to demonstrate optical behaviour. Like ibn Sina, ibn al-Haytham was prolific, writing over 200 treatises and through which he became known to European scholars as Alhazen.

In addition to the social and philosophical space created for natural philosophy by the physicians, Islamic scholars also gained a powerful new tool in the form of an improved mathematical system. The tool was Hindu-Arabic numerals and placeholder mathematics. Originally an import from India, it profoundly changed Islamic scholarship, opening up new classes of problems and methods of calculation. It was pioneered by Muhammad ibn Musa al-Khwarizmi (c. 780–c. 850) in a work called *Concerning the Hindu Art of Reckoning*, which, in addition to the symbol set that was the precursor to the modern notation system, introduced zero as a mathematical object. While the Greeks had understood the concept of nothing, they had explicitly rejected "nothing" as a mathematical term, and it was not a necessary concept for geometry.

Al-Khwarizmi went on to produce *Al-jabr wa'l muqabalah*, which became known in the West as *Algebra*; we get the term "algorithm" from his name. It was from this work that solving for unknowns was developed. Al-Khwarizmi also demonstrated solutions for various quadratic equations including the use of square roots. Historians have argued over whether he was an original thinker or was only a compiler of earlier work such as parts of Euclid's *Elements* and Ptolemy's *Almagest*. Although the answer cannot be definitive unless new material is found, clues such as the fact that al-Khwarizmi's calculation of coordinates for locations were more accurate than those of Ptolemy suggest an intellect capable of difficult and exacting work.

The greatest Islamic natural philosopher of the era was Abu Arrayhan Muhammad ibn Ahmad al-Biruni (973–1048). A polymath by any standard, al-Biruni's studies covered astronomy, physics, geography and cartography, history, law, languages (he mastered Greek, Syriac, and Sanskrit and translated Indian manuscripts into Arabic), medicine, astrology, mathematics, grammar, and philosophy. He was taken by the ruler Mahmud (whether as a guest or prisoner is unclear) to India, where he composed *India*, a massive work that covered social, geographic, and intellectual aspects of Indian culture. He corresponded with ibn Sina and became known as al-Ustadh, meaning "Master" or "Professor." Among his accomplishments were calculating the radius of the Earth, finding that it was 6,339.6 kilometres (extremely close to the modern value); making detailed observations of a solar and lunar eclipse; and writing about the use of mathematics and instruments in his work *Shadows*.

Geography and cartography are probably the branches of science most likely to draw on knowledge and traditions of a number of different cultures and societies, since they require extensive travel or knowledge of other parts of the world. The development of Islamic cartography and geography demonstrate the complex interconnections among different knowledge communities, as information was communicated, appropriated, and adapted for use by the growing Islamic empires.

The earliest cartographic traditions in this area were an amalgam of pre-Islamic Arabian, Persian, and Indian influences. The earliest mapping was done during the Abbasid rule in Baghdad after 762, where the rulers encouraged science and literature and recognized that the conquered nations such as the Sassanids and Byzantines had much to offer. Indian knowledge was also seen as important and transmitted to the court through traders and scholars. Early geographical work owed much to the Indian traditions, especially placing the prime meridian at Ujjain (Arin), and seeing Lanka (Sri Lanka) as the "Cupola of the Earth" (the central point of the inhabitable world). From the Persians, geographers took up the concept that the inhabited world was divided into seven *kishrars* or regions, with six regions encircling the central one of the Iranian area, in a sort of lotus flower image.

Under the Caliph al-Ma'mun (r. 813–833), mapping began to develop. Al-Khwarizmi produced tables of longitude and latitude coordinates for places, influenced by Ptolemy's *Geographia*, which was translated during the

Alchemy

Islamic scholars were not content to confine their understanding of the world to a philosophical system. They wanted to utilize their knowledge, and the greatest exploration of the application of philosophy to the material world was in the study of alchemy. The etymology of the name encapsulates the intellectual heritage of the study. The origin of the term probably came from the ancient Egyptian *khem*, meaning black. The Greek *khēmia* meant "art of transmutations practised by the Egyptians," since Egypt was the Black Land. In Arabic, the Greek root was transformed into *al-kimiyā*, meaning "the art of transmutation" and hence from Arabic into Latin and English.

caliphate by Abu Yusuf Ya'qub ibn 'Ishaq aṣ-Ṣabbaḥ al-Kindi (c. 801–c. 873). Thus, Greek ideas started to interact with the earlier Persian and Indian views on the placement of landmasses and inhabited places on the earth. While Muslim geographers rejected Ptolemy's map projections (based on a grid of latitude and longitude coordinates), they were interested in establishing the coordinates for particular places. They were also able to correct Ptolemy's length of the Mediterranean. A deeply scholarly debate about the location of the prime meridian developed since Ptolemy had placed it at the Fortunate Isles in the west, in contrast to the Indian placement of the prime meridian in the east at Ujjain.

The earliest Islamic maps that still exist today came from a separate tradition, that of the Balkhi school of geographers of the tenth and eleventh centuries. These maps were based on knowledge from travel, trade routes, and postal routes of the far-flung empire. The cartographers were themselves extensive travellers, many from the western caliphates of Egypt and Palestine. For example, Abu

al-Qasim Muhammad ibn Hawqal (travelled 943–969) was born in Upper Mesopotamia and spent his life travelling through Islamic Africa, Persia, Turkestan, and Sicily. In this way, practical experience was as important to geographical knowledge as the scholarly traditions of the earlier geographers.

By the mid-eleventh century, Abu Rayan al-Biruni was creating another tradition of Islamic geography and cartography. A prolific translator and mathematician, Abu Rayan al-Biruni became interested in the mathematical aspect of geography and cartography. He combined a strong knowledge and understanding of both Greek and Indian sources, as well as the work of the Balkhi school, to develop some new theories about the Earth. He remeasured a degree of latitude and tried (somewhat unsuccess-fully) to measure the difference in longitude between locations. He developed a method to determine the direction of Mecca from any location, indicating the interconnection of geographical investigation and social and religious life. Mapping the Muslim world was only possible with these sorts of interactions among knowledge communities both east and west.

Alchemy was in some ways the precursor to the modern material sciences of pharmacology (iatrochemistry), chemistry, mining and smelting, and parts of physics and engineering, as well as aspects of biology such as the study of fermentation, decay, and reproduction. At a basic level, alchemists were trying to identify, classify, and systematically produce useful or interesting substances. Yet this aspect of alchemy, which may seem eminently useful and complete to us, was regarded as mere craft and not the objective of the study at all. The true study of alchemy was the manipulation of the material world, particularly the transformation of substances from one kind to another. It was in this study that alchemists ventured into a mystical realm that had spiritual and religious implications.

The transformation of material is in many ways an everyday occurrence. Wood turns into flame, ice turns into water, and seeds turn into plants. Some transformations seem more magical than others; smelting, for example, takes hunks of rock and transforms them into metal. All societies that manipulated materials developed systems of explanation that covered both the process and the reason materials could be transformed. These explanations were often kept secret not only for trade and safety reasons but also because powerful supernatural forces were involved and so involved religious concerns as well. Thus, alchemy developed both an exoteric, or public, aspect and an esoteric, or secret, element.

Islamic alchemy was founded on Egyptian and Greek ideas about the material world. Through the Egyptian connections came Hermeticism, from Hermes, the Greek name for the Egyptian deity Thoth, the father of book learning and creator of writing. Hermeticism was a blend of Egyptian religion, Babylonian astrology, Platonism, and Stoic thought. The Hermetic documents were likely compiled in the second century BCE and had strong occult aspects. To round out the spiritual side, alchemy was also affected by Gnosticism, which started in Babylon and influenced early Christianity. The Gnostics were strong dualists who saw the world in terms of antagonistic pairs such as good and evil, light and dark. Knowledge of certain things could be gained only by "gnosis," or enlightenment that came from inner awareness rather than reason or study. Both Hermetic and Gnostic studies acquired a heritage of secrecy because of potential conflicts with more powerful religions and the desire of adherents to guard their esoteric knowledge.

From the Greeks came the Aristotelian description of matter combined with neo-Platonic concepts of the Ideal. In addition to Aristotle's division of matter, in his *Meteorologica* (which discusses the condition of the terrestrial realm) the earth is described as a kind of womb in which metals grow. Less perfect or base metals such as lead have a natural inclination to become noble metals, seeking ultimately the perfection of gold if the conditions were right. This was linked to both Aristotelian and Platonic ideas about the original source of differentiated matter (the four elements) from a single undifferentiated prime matter. Prime matter had no "pattern," so the alchemists thought it could be made to take on the pattern of terrestrial matter. The key to this transmutation process was often thought to be a kind of catalyst. This agent was known by a number of names, but the most common was the "Philosopher's Stone," which was mentioned as early as 300 CE in the alchemical collection *Cheirokmeta* attributed to Zosimos (fl. 300 CE). Whether the Philosopher's Stone was an actual object, the product of alchemical processes, or a spiritual state depended on the theory of the alchemist.

It is also from Zosimos that we learn of one of the earliest women to practise alchemical research: Miriam, sister of Moses (c. third century BCE), later known as Maria the Jewess, although it is not certain that she was Jewish and she was not the sister of the biblical Moses. Miriam lived in Alexandria and was interested in chemical processes. Zosimos attributes to her the creation of a high-temperature double-boiler for experiments using sulphur and other equipment, and her name survives to the modern age in the French term *bain-marie*, referring to a double-boiler in cooking.

The intertwining of practical skills, religious and mystical thought, and philosophy plus the secrecy of the practitioners makes alchemy a difficult practice to trace or understand. Greek material from the earlier period is not extensive and mostly practical, dealing with dyeing, smelting, and pharmacology. One of the interesting connections that does survive is the association of the planets with various metals.

As bits and pieces of Greek natural philosophy were disseminated through the Arabic-speaking world, the texts on the nature and structure of the material world hinted at the ability to manipulate it. The beauty of secret knowledge is that it makes all things possible, so the lack of clear antecedents, rather than hindering the interest in alchemy, actually spurred its creation among Islamic thinkers. One of the greatest sources for both Islamic and later European alchemy was the work attributed to Jābir ibn Hayyān. His dates are uncertain, but most likely c. 722 to c. 815. While there may have been a real person with that name, it is clear that the majority of work ascribed to him was compiled by the Ism'iliya, a tenth-century Muslim sect; it is not certain which, if any, texts were written by him.

Over 2,000 pieces of text have been attributed to Jabir ibn Hayyan, most of them of much later production, but the *Books of Balances* and the *Summa Perfectionis* (in its Latin form) cover the central aspects of his alchemy. Jabir starts from an Aristotelian foundation, accepting the four elements and the four qualities, but extends Aristotle's idea of *minima naturalia*, or smallest natural particles, as the basis for the difference between metals. The more densely packed the particles, the denser and heavier the metal. The objective of the alchemist was to transform less noble metals into gold by manipulating the structure and packing of the particles by a process of grinding, purification, and sublimation. The process was also governed by mercurial agents that were either catalysts or active components in the process of change. These agents were referred to by Jābir as medicines, elixirs, or tinctures, which reinforced the biological model of metals—the purification of metal was seen as akin to curing disease or purification of the body.

○ or ☉	Gold [Sun]
☽	Silver [Moon]
♀	Copper [Venus]
♂	Iron [Mars]
☿	Mercury
♄	Lead [Saturn]
♃	Tin [Jupiter]

♈	01. Calcination	*Aries, the Ram*
♉	02. Congelation	*Taurus, the Bull*
♊	03. Fixation	*Gemini, the Twins*
♋	04. Solution	*Cancer, the Crab*
♌	05. Digestion	*Leo, the Lion*
♍	06. Distillation	*Virgo, the Virgin*
♎	07. Sublimation	*Libra, the Scales*
♏	08. Separation	*Scorpio, the Scorpion*
♐	09. Ceration	*Sagittarius, the Archer*
♑	10. Fermentation	*Capricornus, the Goat*
♒	11. Multiplication	*Aquarius, the Water-carrier*
♓	12. Projection	*Pisces, the Fishes*

2.9 ALCHEMICAL SYMBOLS
The alchemical symbols linked the material world with the universe by assigning each element an astrological symbol and each operation a sign from the zodiac.

While Jabir's work (or that attributed to him) was very influential, especially in the Latin West where he was known as Geber, he was something of an anomaly among Islamic scholars because of his concentration on alchemy. More typical of those scholars engaging in alchemical work was Abu Bakr Mohammad Ibn Zakariya al-Razi (c. 841–925). Trained in music, mathematics, and philosophy and likely able to read Greek, al-Razi became a famous and sought-after physician. He was head of the Royal Hospital at Ray (near modern Tehran) and then moved to Baghdad where he was in charge of its famous Muqtadari Hospital. As a physician, he wrote *Kitab al-Hawi fi al-tibb* (*The Comprehensive Book on Medicine*), a massive 20-volume work that covered all of Greco-Roman and Islamic medicine, and *al-Judari wal Hasabah* (*Treatise on Smallpox and Measles*) that contained the first known description of chicken pox and smallpox.

For al-Razi, alchemy was less esoteric than it was for Jabir, and certain aspects of his work, such as the development of drugs and the use of opium as an anaesthetic, can be seen as an extension of his medical work. His most important alchemical text, *Secret of Secrets* or the *Book of Secrets*, does not, despite its title, reveal the secret of transmutation of base metals into gold. Rather, it is one of the first laboratory manuals. Divided into three sections, *Secret of Secrets* covers substances (chemicals, minerals, and other substances), apparatus, and recipes.

The list of equipment was extensive, covering beakers, flasks and jugs, lamps and furnaces, hammers, tongs, mortars and pestles, alembics (stills), sand and water baths, filters, measuring vessels, and funnels. This list remained quite literally identical to the standard equipment found in alchemical, chemistry, pharmaceutical, and metallurgical laboratories until the middle of the nineteenth century, and most of it is still familiar to chemists even today.

Although the *Secret of Secrets* did not offer a specific method of transmutation, it suggested strongly that it could be done. Al-Razi believed in transmutation and subscribed to the same general alchemical theory proposed by Jabir. What separates the two is al-Razi's concentration on practical issues and systematic approach. (See figure 2.10.) For al-Razi, alchemy developed from experience working with materials, rather than from a body of theory that presupposed chemical behaviour. Because of the practical advice he offered, his work became extremely popular in the Latin West, where he was known as Rhazes.

Astronomy in Islam

The stars had guided trade caravans from before the time of Muhammad, and astrology (developed by the Persian Zoroastrians) was important to Abbasid leadership, ensuring that astrologers had a high status in the courts of the early Islamic rulers. In addition to these uses for stellar observation, the injunction that the faithful should pray toward the Ka'bah in Mecca added a particular requirement that engaged astronomers and geographers (often, as in the case of Ptolemy, the same person) in a long and detailed program of

2.10 TABLE OF SUBSTANCES ACCORDING TO AL-RAZI IN *SECRET OF SECRETS*

MINERAL	VEGETABLE	ANIMAL	DERIVATIVE
(see chart below)	Little used	Hair	Litharge (yellow lead)
		Bone	Red lead
		Bile	Burnt copper
		Blood	Cinnabar
		Milk	White arsenic
		Urine	Caustic soda
		Egg	...
		Mother of pearl	Etc.
		Horn	
		...	
		Etc.	

TABLE OF MINERALS					
SPIRITS	BODIES	STONES	VITRIOLS	BORACES	SALTS
Mercury	Gold	Pyrite	Black	Bread borax	Sweet
Sal	Silver	Tutia	White	Natron	Bitter
Ammoniac	Copper	Azurite	Yellow	Goldsmith's boax	Soda
Orpiment	Iron	Malachite	Green		Salt of urine
Realgar	Tin	Turquoise	Red	...	Slaked lime
Sulphur	Lead	Haematite		Etc.	Salt of oak
	Chinese lead	White arsenic			(Potash)
		Kohl			...
		Mica			Etc.
		Gypsum			
		Glass			

1. These tables come from al-Razi, *Secret of Secrets*, in *Alchemy*, ed. E.J. Holmyard (New York: Dover, 1990) 90.

observation. Islamic astronomy was also necessary for timekeeping because the Q'ran mandated the use of the lunar calendar for all religious activities.

One of the first Islamic leaders to support astronomical work was the Abbasid Caliph al-Ma'mun in the ninth century. This helped to give astronomy a level of prestige that continued through the Golden Age. The first significant Arabic work on astronomy was *Zij al-Sindh* by al-Khwarizmi in 830. It was based primarily on Ptolemaic ideas, setting the theoretical framework for later astronomers, but it also marks the beginning of independent work in the Islamic world.

In 850 Abu ibn Kathir al-Farghani wrote *Kitab fi Jawani* (*A Compendium of the Science of the Stars*) that extended the Ptolemaic system introduced by al-Khwarizmi, corrected some of the material, and included calculations for the precession of the Sun and the Moon as well as a measurement of the circumference of the Earth.

The widespread interest in astronomy also led to the development of astronomical instruments. Although the astrolabe was well known to Greek astronomers, the technical skills of Islamic craftsmen led to the creation of very good astrolabes. One of the earliest surviving examples was made by Mohammad al-Fazari around 928. Using a variety of sundials, quadrants, and armillary spheres, Islamic astronomers compiled extensive star catalogues.

Although there were observatories in many of the major cities in the Islamic world, the most influential was established by Hulagu Khan in the thirteenth century at the city of Maragha. Its construction was overseen by the great Persian polymath Nasir al-Din al-Tusi (1201–74). In addition to his many scientific works, he identified the Milky Way as a collection of stars, an observation not confirmed in the West until the work of Galileo. Tusi is also famous for creating what is called the Tusi-couple, which places a small circle within a larger one so that a point on the small circle will oscillate in a regular fashion as the two circles rotate around their common centre. This mathematical device allowed Tusi to remove Ptolemy's awkward equant from astronomical calculations.

During the thirteenth and fourteenth centuries, astronomers following Tusi's lead were able to eliminate most of the extra motions associated with Ptolemy's schema, with the exception of the epicycles. Tusi's student, Qutb al-Din al-Shirazi (1236–1311), worked on the problem of Mercury's motion. Later, Ala al-Din ibn al-Shatir (1304–75), who worked as the religious timekeeper at the Great Mosque of Damascus, found a way to represent the motion of the moon. When Copernicus began his work, which would transform the model of the heavens by placing the sun at its centre, he appears to have had access to both Tusi's and al-Shatir's work, showing how instrumental Islamic astronomers were to the development of astronomy worldwide.

On the Heavens and Number around the Globe

The desire to understand nature through mathematics and astronomy has been a human impulse seen in almost every civilization. We can trace the development of these skills in the celestial observations left by people in the Americas, Australia, and the Pacific Islands. In particular, we have records from the Maya and the Aztec showing that they recorded the motion of the heavens and developed the mathematics needed to create calendars. These observations were primarily done for religious reasons, but they were also part of the practical planning of activities such as planting and harvesting.

The Mayan civilization built across the Yucatán Peninsula was ancient, beginning around 8000 BCE, but the age of greatest intellectual activity was during the Classical period from 250 CE to 950 CE. The Maya had good astonomers and mathematicians. Much of their work was done for religious purposes, but they left a record of significant mathematical and astronomical insight. The Maya recorded the motion of the sun, moon, Mercury, Venus, Mars, Jupiter, and many stars. Their system was geocentric and their observations allowed them to predict eclipses and chart the future position of stars, in many cases with greater precision than European astronomers of the period. They had a special interest in Venus, calculating its 584-day cycle with great accuracy. This may have been because Venus was astrologically associated with war and change.

The Maya created two calendars. The first was a solar count of 20-day periods known as *winal*. A year consisted of 18 *winal* plus a five-day period called the *wayeb*. This period was considered dangerous, when the division between natural and supernatural realms was opened. The Maya projected their calendar far into the future, calculating a span of 63 million years, although in practical terms the longest unit was the *ba'k'tun*, which recorded a period of 394 years. The second *tzolk'in* calendar was a 260-day cycle and was used for religious rituals. It is a subject of much debate about why the *tzolk'in* had 260 days, since this does not match up with any astronomical period. It may have been a numerological construct (13 × 20 for example) or even a measure of the human gestation period. Whatever the reason, the two calendars nonetheless spread throughout Mesoamerica.

According to one correlation between the Mayan *bak'tun* and the European Gregorian calendar, a *ba'k'tun* ended on December 20, 2012. This occurrence was taken by some to be a prophecy of apocalypse and was used as a plot element in the Hollywood movie *2012*. A number of pseudo-scientific documentaries such as

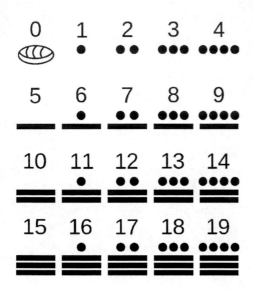

2.11 MAYAN NUMERALS

2012 Apocalypse presented the end of the *ba'k'tun* as a Mayan prophecy of doom. These shows had no foundation in history or science and are part of the problem with misuse of science. In reality, the end of the *ba'k'tun* was simply followed by the start of the next.

Mayan calendars were in part so accurate because the Mayan civilization had a good mathematical system. The Maya used a placeholder vigesimal system (base 20) and had a symbol for zero. They used a series of dots and bars to write their numbers, making basic calculations easy. (See figure 2.11.) In many ways, the Maya were more advanced than their European or Asian counterparts when it came to number theory. What has not been discovered is any systematic use of geometry. Although Mayan architecture makes it clear that they could create a variety of angles, including a consistent 90° angle (probably using knotted ropes) and could align structures to the cardinal directions, we have not discovered any indication that they had developed theoretical principles of geometry.

The Aztec empire began around 1426 and centred on the city of Tenochtitlan (now Mexico City), north of the region controlled by the Maya. The Aztec empire lasted until the Spanish conquest in 1520. Given their proximity to the Mayan world, it is not surprising that the Aztecs had a similar affinity for mathematics and astronomy. Astronomy was so important that it was part of formal education at the *calmecac* schools, along with writing and theology. The Aztecs tracked the motions of the stars and planets and used the observations in the construction of temples and buildings. The best-known example of this is the Templo Mayor, oriented so that on the March 21 equinox, the sun was observed between the Tlaloc and Huitzilopochtli shrines. The Aztec astronomer-priests used the same two calendar systems as the Maya. One of the most important Aztec artefacts is the Calendar Stone (or Sun Stone). (See figure 2.12.) Carved c. 1479, it shows both calendars and a host of religious symbols.

Aztec mathematics borrowed ideas from the Maya and Olmec people. Like the Maya, the Aztecs had a vigesimal system and used dots and bars to represent numbers, adding other symbols for larger numbers. Unlike the Maya, the Aztecs do not appear to have used a symbol for zero, but understood the concept. Mathematics was used for calendars, but also for taxes based on land area of farms and surveying. The Aztec language includes words for tools such as a plumb line and a level.

There was no contact between the Americas and the rest of the world until the Europeans arrived in the late fifteenth century. Their ideas do not appear to have influenced other civilizations, but the history of this region tells us about the widespread interest in mathematics and astronomy. Anywhere people could look up at the sky over a period of time, they recorded and exploited their observations.

In the European context, Western historians have become increasingly aware that international contacts in the first centuries CE were greater than previously imagined. While contacts between the West and the Islamic world were direct, the vast distances and often rugged terrain that separated Europe and the Middle East from Asia were thought to have precluded or seriously limited contact between the two regions until the spread of Islam. It is now clear that this was not the case. For example, when Alexander the Great took his army east into India in 326 BCE, he already knew that great cities and empires existed on the subcontinent and beyond. This raises a complex question about the origins of Western science, since the ideas that have been associated with European natural philosophy may have been influenced by ideas from other parts of the world and in turn may have influenced other cultures. We can no longer claim that the origin of science was an independent product of the West; rather, it involved complex and long-standing trade in ideas and technologies with parts of the world far away from Greece and Rome.

We know that a host of inventions including gunpowder, paper, and the spinning wheel made their way from China to Europe, while from India came Hindu numbers (which we also call Arabic numbers in recognition of their path through the Muslim world), placeholder mathematics, and wootz steel, better known in the West as

2.12 AZTEC CALENDAR STONE

Damascus steel (again because the contact for Europeans was the Islamic world). What is less clear is whether particular explanations and discoveries were transmitted in a similar way, or whether they were arrived at independently in each culture. For example, every society that recorded the motion of the sun and moon independently discovered the solar and lunar calendars, and all societies interested in mathematics developed some version of the Pythagorean theorem.

The rise of the empires of the East was predicated on the same developments as those of the Mediterranean basin: agriculture, job specialization, bureaucracy, and urbanization. In India and China, natural philosophy became a part of the intellectual heritage of those regions, with Indian scholars more closely linking their observations of nature to the Vedic texts that were the foundational religious and social texts for many of the people of the region, while Chinese and Far Eastern scholars tended to be more pragmatic, in keeping with the less supernatural ideas of Taoism. Asian scholars were also influenced by Buddhism as it spread after the fifth century BCE.

The Vedic tradition in India came from a series of texts written between about 1500 and 900 BCE. These texts, while primarily religious, also include material about mathematics, geometry, biology, and medicine. Vedic mathematics was developed as part of the methods for correctly performing rituals and was studied as part of the six disciplines of the *Vedangās* (*Ancillaries of the Vedas*) starting around the sixth century BCE, especially *kalpa* (rituals) and *jyotiṣa* (astrology). Mantras from this period also showed an interest in large numbers, with some mantras naming units up to a trillion.

The *Baudhayana Sulba Sutra* by Baudhayana (fl. eighth century BCE) was an early mathematical text that identified what we would call the Pythagorean relationship and gave some of the common whole-number triplets (3, 4, 5 and 7, 12, 13). It also gave a formula for the square root of two, indicating that the mathematicians of the time understood that the most basic solution to the relationship of the sides of a right triangle to the hypotenuse could not be expressed as an integer.

One of the greatest gifts to mathematics within the Vedic tradition was the invention of positional notation using numerals. The origin of this system is not clear, but by 499 CE the astronomer and mathematician Aryabhata was using placeholders, although he used letters rather than numerals. Numerals much closer to modern symbols came into use around 600 CE. This system of notation was known to Syrian scholars by 662 CE and was later made much more common when they were adopted by Arabic scholars such as Al-Khwarizmi (which is why we now call them Arabic numerals). The first European mention of the new number system was in the second *Codex Vigilanus*, a compilation of different writings, completed in 976 CE.

The earliest known Indian astronomical text was the *Vedānga Jyotişa*, composed sometime between the sixth and fourth centuries BCE. Although it was religious in nature and created in part to regulate religious observance, it contained practical information on solar and lunar cycles, a list of planets, and guidance for celestial observation. Aryabhata argued that the Earth rotated, while still placing it in the centre of the universe in a geocentric model.

One of the most intriguing natural philosophical ideas came from the Hindu scholar Kanada (fl. second century BCE, although dates as early as sixth century BCE have been proposed). Kanada was interested in matter theory, studying a form of alchemy known as *rasavādam*. In part of his work, he argued for the existence of atoms, which he described as indivisible, indestructible, and eternal. Some modern scholars have suggested that Kanada's atomism, while more abstract than the Greco-Roman atomism of Democritus, was more complete.

Although Indian natural philosophers undertook a wide range of investigations and achieved many notable insights, they did not separate natural and spiritual studies of nature, as the Greeks did. Political, religious, and military turmoil such as the Arab conquest of the Sindh c. 712 may have disrupted the development of Indian natural philosophy in the sub-continent. Once India was part of the Muslim world, Islamic natural philosophy (borrowing useful concepts from the Indians) was supported by the rulers.

An important bridge between China, India, and the Mediterranean world was created around 263 BCE when Ashoka the Great (304–232 BCE), who controlled most of modern-day India, Afghanistan, Pakistan, and Bangladesh, converted to Buddhism. Ashoka sent emissaries to neighbouring regions as far away as Alexandria in the west and Burma in the east. Buddhism likely reached China around 70 CE, and opened a wider exchange of ideas between East and West. Some modern scholars have argued that Indian mathematics in particular was influenced by Chinese work, while Greek natural philosophy flowed into India through the remnants of the Greco-Bactrian kingdom established by Alexander the Great.

Natural philosophy in China represents a challenge to modern historians and philosophers of science. A massive 24-volume history of Chinese science and technology entitled *Science and Civilization in China* (1954–2004) was undertaken by the scholars Joseph Needham (1900–95) and Wang Ling (1917–94), so the vast range and depth of knowledge and invention are well known. China was, for most of its early history, the richest and most technologically advanced empire in the world. Its scholars were highly educated and commanded vast resources compared to European or even Islamic scholars. Several specific areas of research, such as alchemy and astronomy, were extremely well developed, but despite these advantages, a unified natural philosophical system did not develop.

Alchemy in China was closely tied to Taoist ideas and medicine, and there was no clear distinction between alchemical work and what we would call pharmacology. The study of alchemy extends back to at least the second century BCE, since concern about alchemists existed from at least 144 BCE, when the emperor issued an edict outlawing the making of "counterfeit" gold on pain of death. Although there was much work done on transmutation, the primary focus was on immortality. One of the oldest alchemical texts was *Tsan-tung-chi* (authorship unknown), that appeared around the third or late second century BCE. It described the way to make a golden pill that would make a person immortal. The noted Taoist scholar and high government official Ko-Hung (also Ge Hong, 283–343 CE) wrote extensively about alchemy and immortality. His general theory was based on the purification and transmutation of metals as a way to remove those negative aspects of biology that caused aging. Chinese alchemists, like Western alchemists, were in a difficult position, having to reveal enough to establish their skills, while needing to keep specific details of their work secret, both as trade secrets and to protect people from the dangers (spiritual and physical) of alchemy. While a strong *materia medica* developed, alchemy never became a comprehensive study of matter in China.

Chinese cosmology was a more unified study than alchemy, but it was also tied very closely to Taoist ideas about existence and the place of things in the universe. Early Chinese astronomers identified the solar and lunar calendars and plotted the paths of the visible planets. They were good observers and noted the passage of comets as well as the appearance of a supernova in 1054. There was a surge of astronomical work during the Han dynasty (25–220 CE), and then again during the Tang dynasty (618–907 CE) when an influx of Indian astronomers arrived in China with the spread of Buddhism.

Chinese observers were particularly adept at creating star catalogues, and more than 2,000 stars were listed by Zhang Heng (78–139 CE), who also calculated solar and lunar eclipses. The information from these catalogues was also used to create some very detailed armillary spheres, and the astronomer Su Song (1020–1101) and his colleagues created a massive mechanical clock tower in 1092 that included a moving armillary sphere and a celestial globe. (See figure 2.13.)

After the establishment of the Mongol Empire, Chinese astronomers worked more closely with Islamic scholars. Kublai Khan (1215–94) brought Iranian astronomers to

Beijing to build an observatory and started a school of astronomy around 1227. A series of observatories were built in Beijing, and the one completed in 1442 has been preserved and is one of the oldest pre-telescopic observatories still in existence. Through these contacts with Islamic astronomers, the Chinese learned about the Ptolemaic system.

By the sixteenth century, the Chinese invited a number of Europeans, particularly Jesuits, to teach them European natural philosophy and Ptolemaic astronomy. Although there was debate among Chinese astronomers about the Ptolemaic model, most of them rejected it because it required a material substance to occupy space, while their most widely held view was of celestial objects in an infinite void. As was the case with Indian science, Chinese science was powerful in particular subjects but never developed an overarching explanatory model or method that was not religious in nature.

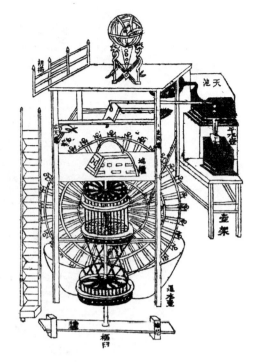

2.13 CHINESE MECHANICAL CLOCK (1092)

Chinese mechanical clock. From *Hsin I Hsiang Fa Yao*, ch. 3, p. 4a.

Conclusion: The End of the Islamic Renaissance

One of the last great thinkers of the Islamic Renaissance was Abul-Waleed Muhammad ibn Rushd (1126–98), who was known in both the Arabic and the Latin worlds as the Great Commentator or simply the Commentator. Rushd was educated in philosophy and trained as a physician, but he worked primarily as a judge and expert in jurisprudence and lived most of his life in Spain at Córdoba. In many ways Rushd represents both the power and the waning of power in Islamic natural philosophy. His commentaries on Aristotle were not based on primary sources but rather on Arabic translations, so he was not attempting to return to the original material. He wrote three sets of commentaries: the *Jami*, a simplified overview; the *Talkhis*, an intermediate commentary with more critical material; and the *Tafsir*, which represented an advanced study of Aristotelian thought in a Muslim context. These were fashioned as educational steps to take the novice from an introduction to an advanced understanding of Aristotle and, in effect, created an Islamic Aristotle.

What Rushd added to natural philosophy was not original work on nature, but a powerful synthesis that represented a well-established intellectual heritage. He presented the most dedicated version of Aristotelianism, essentially arguing for the perfection of Aristotle's system of logic and philosophy. From this position, Rushd argued that there were two kinds of knowledge of truth. The knowledge of the truth that came from religion was based on faith and thus could not be tested. It was the path to truth for the masses, since it did not require training to understand because it taught by signs and symbols. Philosophy, on the other hand, presented the truth directly to the mind and was reserved for an elite few who had the intellectual capacity to undertake its study. This did not mean that religion and philosophy conflicted but that they could *not* conflict. Philosophy might be an intellectually superior way of understanding truth, but it did not make truth. It followed that any truth revealed by religion would be the same as the truth arrived at by philosophy.

This support for philosophy and the declaration of the relation of philosophy and religion had a profound influence on medieval European scholars, particularly Thomas Aquinas (1225–74). Known as Averroes to the Latins, Rushd's work became a key component for much of their work on Aristotle, especially for a group called the scholastics. Rushd was both a source for medieval Aristotelianism through the commentaries (especially before Aristotle's work itself was widely available) and a philosophical anchor. His support for the intellectual superiority of philosophy also made his work the target of criticism, and supporters were accused of heresy or even atheism. The position Rushd argued has been echoed in considerations of the relationship between science and religion down to the modern day. For many natural philosophers and scientists the idea that the study of nature (philosophy) could reveal the truth of God's creation was not merely a justification for the reconciliation of reason and faith but a call to pursue the study of nature.

The interest in natural philosophy that grew during the Islamic Renaissance faded as the Islamic world became fractured and in general more conservative. The *hakim* were often brilliant individual thinkers, but for the most part they failed to reach a kind of critical mass that would make research in natural philosophy a desirable commodity. The natural philosophers were also victims of their own success, for, having created a model of ideal Arabic natural philosophy (particularly in areas such as medicine and astronomy), the schools slowly shifted from active research to perpetuation of established work. The failure to create an enduring research ethic was the result of a number of factors such as the political turmoil that disrupted all aspects of society and the swings from a high level of religious tolerance to strict fundamentalism that occurred almost overnight when leadership changed hands, making it potentially dangerous to engage in work that suddenly might be deemed unacceptable. It may also have reflected the level of respect accorded to the work of the great thinkers, which made new work more difficult to disseminate as the old work increasingly came to be seen as orthodoxy.

Cultural and religious changes also affected the place of natural philosophy. Mysticism on the one hand and more doctrinaire Islam on the other rose as the dominant religious forces in the thirteenth and fourteenth centuries. The Islamic Empire was increasingly under threat militarily, with incursions by the Mongols in the east, the reconquest of Spain in the west, and infighting among the kingdoms within the empire itself. Islamic law increasingly defined the proper sphere for human creativity; this did not include skepticism or any real place for personal opinion or secular corporate identity. In parts of the Muslim world pictures of people or nature were banned because it was thought they were idolatrous. This placed rather severe limits on certain kinds of investigations such as botany and hampered the communication of observations through texts. The rulers of religious states were also concerned that philosophy of any kind would conflict with theology, and so they were less willing to support work by scholars interested in those topics. There may also have been an element of psychological superiority that came from the power and glory of the richest Islamic states. In the early days of Islam, Greek knowledge and Roman power were still part of common knowledge, but 500 years after the fall of Rome and the end of the Byzantine Empire, the old world had clearly been surpassed by the new. Why then waste time and effort studying the remnants of a failed (and pagan) society?

Even the appearance of the barbarous and ill-educated knights of Western Europe seemed of little threat to the power of the Islamic world.

Essay Questions

1. What problems did Ptolemy's system solve and what problems were left unsolved?

2. Why do we consider Galen an Aristotelian?

3. How did natural philosophy develop in the Islamic world? Were Islamic scholars important innovators?

4. What topics were most important to Chinese natural philosophers and why did their study of nature develop as it did?

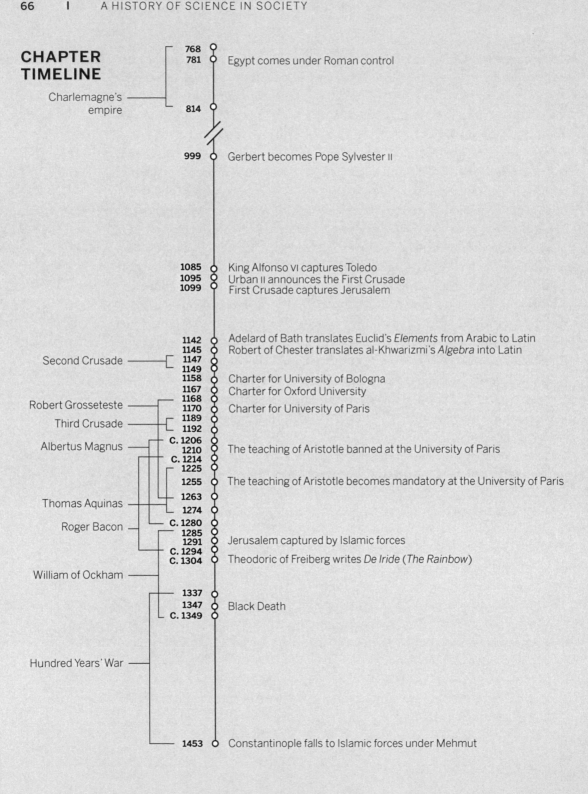

CHAPTER TIMELINE

Charlemagne's empire

768
781 Egypt comes under Roman control

814

999 Gerbert becomes Pope Sylvester II

1085 King Alfonso VI captures Toledo
1095 Urban II announces the First Crusade
1099 First Crusade captures Jerusalem

1142 Adelard of Bath translates Euclid's *Elements* from Arabic to Latin
1145 Robert of Chester translates al-Khwarizmi's *Algebra* into Latin
Second Crusade **1147**
1149
1158 Charter for University of Bologna
1167 Charter for Oxford University
1168
Robert Grosseteste **1170** Charter for University of Paris
1189
Third Crusade **1192**
Albertus Magnus **C. 1206**
1210 The teaching of Aristotle banned at the University of Paris
C. 1214
1225
1255 The teaching of Aristotle becomes mandatory at the University of Paris
Thomas Aquinas **1263**
1274
Roger Bacon **C. 1280**
1285
1291 Jerusalem captured by Islamic forces
C. 1294
C. 1304 Theodoric of Freiberg writes *De Iride* (*The Rainbow*)
William of Ockham

1337
1347 Black Death
C. 1349

Hundred Years' War

1453 Constantinople falls to Islamic forces under Mehmut

THE REVIVAL OF NATURAL PHILOSOPHY IN WESTERN EUROPE

3

Successive waves of invasions following the fall of Rome disrupted all aspects of life in Europe. The physical destruction of war and economic collapse destroyed many collections of texts, educational institutions fell into ruins, and society turned from the pursuit of knowledge and empire to basic survival. Despite the dire conditions, not all ancient knowledge was lost. Greek works survived in the Byzantine Empire, and certain texts remained known in the West, including most of Plato's *Timaeus*, parts of Galen's medical treatises, elements of Ptolemy's astronomy, and some of Boethius's studies in mathematics and astronomy, as well as Aristotle's logic. These resources were valuable but scattered and fragmentary. The best and brightest minds were gathered by the Church and turned their thoughts to questions of theology. Since certain aspects of Christian theology had to deal with issues of the physical world, there continued to be a need for information about the material realm, whether it was astronomy for calendars to keep track of feast days and observances or medicine to meet the Church's duty to care for the ill. In the early days of the Church, there was a struggle between those inclined to intellectual activities and those who favoured a more mystical approach. In the long run, the greater managerial skills of the intellectual wing came to dominate the administration of the Church, and the study of nature was included in Western intellectual practice.

In the Latin West during the course of the Middle Ages the Christian Church succeeded in establishing itself as the authority over intellectual as well as spiritual concerns. Therefore, just as had been the case in Islamic countries, supernatural and spiritual issues became intertwined with natural philosophical ones. Thus, despite the influence of Greek natural philosophy in the medieval West, the study of nature became a battleground for the primacy of natural or supernatural explanations once again. At stake were the questions of who controlled knowledge and who had the ultimate authority over truth claims. The answer was a reinventing and reordering of the intellectual universe, with a separation made between the spiritual and natural (or mystical and rational) which was different than that of the Greeks but similarly powerful. As long as that separation was controlled by the Catholic Church, the result was a well-ordered, carefully moderated series of disputations about nature and humanity's place within it. When the Church began to lose its authority in the sixteenth century, that very separation exploded into a cacophony of multiple voices.

The universities became the dominant and necessary space in the creation of both these careful rules of knowledge and the later tensions. The universities, founded in the twelfth century and beyond, provided space sanctioned by the Church and yet were not completely under the Church's control, since scholars were taught not only the prevailing system of scholasticism, which was focused on understanding the revealed truths of Christianity through rigorous syllogistic logic, but also to contest Greek philosophical ideas and methods of questioning while incorporating them into the powerful system of scholasticism. Competition was built into the system; ironically, those very places created to determine and preserve orthodoxy became the site for alternative natural philosophies in later centuries.

Those who studied nature in the medieval period were as concerned with the method of acquiring knowledge as with its application, and so there developed a complex dialogue concerning utility and practicality. Unlike Muslim scholars, who were interested first in applicable sciences such as medicine and astronomy, Latin scholars were first concerned with the use of natural knowledge as a path to salvation. While some of them explicitly experimented with nature and looked for applications in military, alchemical, and cartographic contexts, others were concerned about the implications of such action. Thus, the utility of natural knowledge, relatively unproblematic for Islamic scholars, was a hard-fought question for Europeans.

During the sixth and seventh centuries Europeans had only limited access to Greek, Roman, and Islamic natural philosophy. By the ninth century growing intellectual activity in Western Europe, particularly in parts of France and the rich Italian city-states, began to support new inquiries. This interest was further spurred as Europeans came into contact with the material and cultural wealth of the Islamic world. Material that Islamic scholars had preserved, commented on, and expanded, especially in the areas of logic and mathematics, medicine, alchemy, astronomy, and optics, increasingly came to the attention of Latin scholars. The "People of the Book," those who shared the Old Testament as a foundational religious document, were officially tolerated by Islamic rulers, and, as a result, Christian and Jewish, Muslim scholars were often able to visit and use the resources held in Islamic territory. Jewish scholars, who had ties with both Europe and the Middle East and were often multilingual, acted as a bridge between cultures. The libraries in Moorish Spain, particularly the one in Córdoba which contained over 400,000 volumes, became centres of education and recovery of the Greek texts that had been lost to the Latin West.

Charlemagne and Education

During the short-lived Carolingian Empire, which lasted only from 768 to 814 during the reign of Charlemagne (742–814), there was both a renewed interest in intellectual activity and the rebirth of the concept of an empire capable of matching the achievements of ancient Rome. Charlemagne claimed the title of Holy Roman Emperor and thereby established a new Roman era, if not exactly a new Roman Empire. His drive to create a European empire had more than a political effect, because it also shaped people's attitude toward the future. The early Middle Ages were tinged with a certain pessimism and a somewhat backward-looking view of society. This came in part from the belief held by many that the world was entering the end days as described in the Bible. Throughout Europe there were, literally, concrete examples that the past was better than the present, as the remains of the power of Rome dotted the landscape. Ruins of aqueducts, roads, and coliseums were a continual reminder of lost power and lost knowledge. Charlemagne's success started people thinking about the possibility of reclaiming the glory of Rome and about a future that might be better than the present. Matching the wonders of Rome required knowing what the Romans had achieved, and so attention turned to the Greco-Roman heritage.

Charlemagne was a superb general, but even more he was an astute politician who recognized that winning an empire was not the same as holding it together. Citizens must be persuaded to believe that they were better off in the empire than on their own, so Charlemagne worked to establish a uniform system of law, organized the military, improved the churches, and created public works. He placed education at the heart of his reforms, attracting Europe's foremost scholars to his court at Aachen (Aix-la-Chapelle) to manage the empire and help create this new culture.

CONNECTIONS

Natural Philosophy
and Education:
Alcuin and the
Rise of Cathedral
Schools

The growth of natural philosophy or science has always required an education system, since the principles necessary for the systematic study of nature are not self-evident and must be taught. Without such an education, knowledge was easily lost, and worse, the methods to acquire knowledge were also lost. The greatest example of this was the period after the fall of Rome, often called the "Dark Ages," when the light of learning nearly disappeared from Western Europe.

In the Ancient world, most formal education was provided by tutors and then only for the wealthiest families. Advanced education at schools like Plato's Academy or Aristotle's Lyceum trained the elite of the elite. The only other sources of education were the temples, some of which taught basic skills in literacy and mathematics to disciples. After the end of the Roman Empire, only a few churches, mosques, and synagogues offered basic education so that a handful of people could read the holy documents.

When Charlemagne became the Holy Roman Emperor in 800 CE, he had a big problem. Many priests were illiterate, so they could not read the Bible and perform the liturgy. The lack of literacy also meant that running an

The most prominent of these scholars was Alcuin (735–804), who had been educated in Ireland and was head of the cathedral school of York. There the monks had developed a curriculum based on a combination of classical training and Christian theology.

In 781 Alcuin met Charlemagne, who asked him to join his court and be his minister of education. Alcuin accepted and, in addition to developing a school system, educated the royal family and acted as a private tutor to the emperor.

Alcuin helped Charlemagne establish cathedral and monastery schools by imperial edict, and in turn these schools produced clerics with increasing levels of literacy and scholarship. Priests were to be literate, and Charlemagne charged the bishops with the responsibility of ensuring literacy and the delivery of proper religious observance, particularly the reading of the liturgy. While in Charlemagne's service, Alcuin was also instrumental in collecting manuscripts and establishing *scriptoria* for the copying and dissemination of the texts.

Alcuin's curriculum provided the foundation for education in Europe for over 1,000 years. His system was based on the study of the seven liberal arts, divided into two sections called the *trivium* and the *quadrivium*. From the Latin *liber* meaning empire was difficult, since everything from long-distance communications to government reports and tax collecting all required literary and mathematical skills that few people possessed.

Charlemagne gathered many scholars at his court at Aachen. He invited Alcuin of York, one of the most learned men in Europe, to join his court as a member of the Palace School and as his personal tutor. Alcuin influenced Charlemagne to enact educational laws for the Church and to require bishops to establish schools. These schools, called cathedral schools because they were housed in the home churches of the bishops, taught the clergy to read and write. The utility of literacy prompted the expansion of schools to monasteries and even lay (non-religious) schools in towns and cities. Monasteries began to copy and preserve texts in *scriptoria*. A number of

cathedral schools went on to become universities, with the University of Paris being the most famous. By the Third Lateran Council in 1179, the spread of literacy and education led to the Scholastic Movement to recapture the knowledge of antiquity.

Without education, natural philosophy would have disappeared completely from Western Europe. At first, the Church focused on practical aspects of natural philosophy, especially medicine (Galen) and astronomy (Ptolemy), but literacy opened the door for a much broader investigation of natural philosophy. Despite Alcuin's best efforts, however, Charlemagne never learned to read. It was said that he slept with books under his pillow in hopes that the knowledge might transfer by proximity (a practice some students are suspected of continuing to the modern day!).

free, the liberal arts served the purpose of educating the free man to be a good citizen, in contrast with the *artes illiberales*, which were studied for economic gain.

Trivium means place where three roads meet, but it also implies a public space. The three subjects of the trivium were logic, grammar, and rhetoric, and mastering these was the essential first step of education. Through clear thinking, clear writing, and correct speech in Latin (the *lingua franca* or universal language of European scholars), a person was prepared to participate in civilization. The *quadrivium* (or four roads) consisted of geometry, arithmetic, astronomy, and music. Music was the branch of mathematics that investigated proportions and harmony, which might include studying singing or playing instruments but was really concerned with the underlying mathematical theory. The two halves of the liberal arts curriculum represented the two ways of understanding the world, first through language and, once that was mastered, through the patterns of the world discernible only through mathematics.

One of the most gifted students to come out of the reformed schools was Gerbert (c. 945–1003). He studied in France and Spain before becoming headmaster of the cathedral school at Reims. He later became the archbishop of Reims, then of Ravenna in Italy. With the patronage of Otto III of Saxony, he was elected Pope Sylvester II in 999. Gerbert was deeply interested in logic and mathematics and was involved in efforts to find, translate into Latin, and copy Greek and Arabic texts on natural philosophy. When he became pope, he set the tone for the whole Church, raising the profile of natural philosophy and reinforcing the intellectual side of theology.

The Crusades and the Founding of the Universities

Although Alcuin and Gerbert established an intellectual tradition in the Church and began to prepare an audience for Greek and Islamic scholarship, they represented a tiny group interested in the still arcane study of philosophy. Churchmen of this period had a complex reaction to natural philosophy. Augustine, one of the most influential Christian thinkers, felt that natural philosophy could be an aid to theology, but revealed knowledge was always superior to discovered knowledge if there was any apparent conflict. Many theologians argued that the study of the natural world at best was irrelevant and at worst impeded one's hope of salvation. To place Greek natural philosophy at the heart of European scholarship required more than a slow acquisition of the ancient works and their Arabic commentaries and additions. What spurred the Europeans to the greatest action was the military struggle first against Islamic expansion and then for control of Jerusalem and the Holy Land. This both changed the culture of Europe and dramatically increased interest in the Greco-Roman world.

The Mediterranean Sea was almost completely under the control of Islamic forces who held Spain, North Africa, the Middle East, and Asia Minor. In 734 CE their western push into Europe stopped when Charles Martel defeated an Islamic army at Poitiers, ending further challenges to Frankish lands beyond the Pyrenees. Eventually Christian forces pushed the Moors out of Spain, starting with the capture of Toledo in 1085 by King Alfonso VI, although the last of the Islamic territory there was not captured until the late fifteenth century.

The expansion of Islam in the east was resisted by the Byzantine Empire, but under successive waves of Islamic forces starting with Suleman, the eastern European region was slowly conquered. Finally in 1453 Mehmut's army defeated the last holdouts in Constantinople, ending the Byzantine Empire. Refugees from the fall of Constantinople brought manuscripts and a knowledge of Greek to Western Europe, adding a second wave of interest in ancient philosophy. Constantinople was renamed Istanbul by the invaders, and from this base on the western side of the Bosphorus Islamic incursions into the west did not end until the second defeat of Ottoman Empire forces at the gates of Vienna in 1683. Many of the modern problems of the Balkans stem from the historical flux and mix of people and religions that long years of warfare brought to the region.

These external threats to Latin Christendom as well as domestic conditions led Pope Urban II to call Christians together for the First Crusade in 1095. Europe had entered a period of stability that left many of the nobility with little to do but fight among themselves. The knights of Europe were more Spartan than Athenian, mostly illiterate and trained from an early age to withstand the rigours of combat and not much else. With little in the way of new land available, the ruling class was under pressure to provide for second and later sons, since frequently little inheritance was left after the rules of primogenitor placed all the family lands in the hands of the eldest son. When Alexius I Comnenus, the Byzantine emperor, called for help against the Seljuk Turks, a crusade seemed a good way of dealing with many issues at once. Emboldened by the success of Alfonso in Spain, Urban believed that the Latin West could come to the aid of the Greek East against a much-feared enemy, while at the same time the largely idle knights could practise their profession far from home. For the nobility there was pious warfare, adventure, and the potential for land and wealth, while for the Church there was the possibility of controlling the Holy Land, conversions, and striking a blow against a competing faith. And for those supplying the Crusaders, there were significant profits.

The first three crusades—1096–99, 1147–49, and 1189–92—had some success from the Crusaders' point of view, with Jerusalem falling to Christian forces in 1099. Although the capture of Jerusalem was symbolic, the territorial gains were never great, and the European hold on the Holy Land was short-lived. What the Europeans really gained was renewed contact with a wider world. In a sense, natural philosophy returned to the Latin West because its people discovered a craving for spices, silk, fine china, ivory, perfume, and a host of exotic luxury items, many of which came from Asia along the Silk Road and through the Middle East. With these goods, they also traded ideas. Although east-west trade had never completely been cut off, the expansion of the trade in luxury items made cities such as Venice and Florence extremely wealthy. That wealth in turn financed the intellectual and artistic boom of the Renaissance, while adding to the wealth of the Arabic world that controlled the trade between Asia, Africa, and Europe. The desire of Europeans, especially those unable to participate in the Mediterranean trade, to avoid the middlemen and trade directly with the East was also the spur to global explorations in later years.

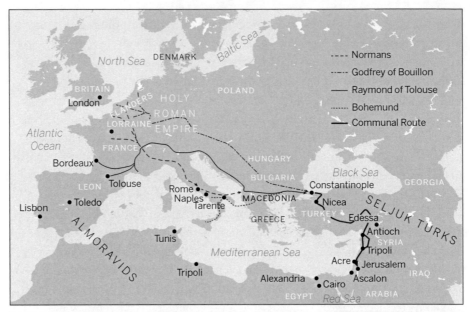

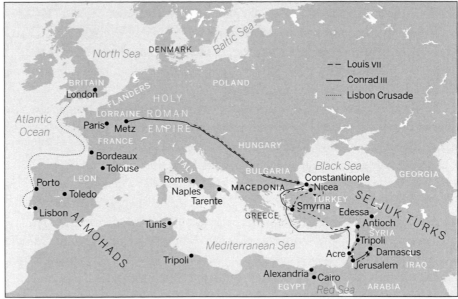

3.1 THE FIRST TWO CRUSADES

This expansion of commerce promoted urbanization, and, in turn, increasing urban populations could support education, including higher education in theology as well as secular topics such as law, the liberal arts, and medicine. Developing in large part out of the cathedral school system established by Charlemagne and in part from models copied from Islamic schools, the first European universities were founded in this period. The University of Paris claims it began in the early 1100s, making it the oldest institution of higher learning in Europe, but by charter the 1158 founding of the University of Bologna is probably the earliest officially organized university. Oxford University was founded in 1167, and the University of Paris was formally established by 1170.

The creation of the universities legitimized the study of natural philosophy and provided a place for scholars to live and work. They became the centres of intellectual debate and the repositories of manuscripts both old and new. As teaching organizations they produced more intellectually rigorous theologians and helped raise the level of literacy among the clergy. They also performed a vital role in training the growing secular managerial class. As well as holding positions of power in Church and government bureaucracies, these literate and university-trained students became essential members of the noble and princely courts.

Just as Islamic scholars had first gathered Greek philosophy and then produced Arabic translations, Western scholars eagerly sought out Arabic manuscripts and set them in Latin. During this period of rapid translation a number of scholars were key in introducing natural philosophy to the Latin audience. Adelard of Bath (c. 1080– c. 1152) undertook a number of translations, concentrating on mathematical texts such as al-Khwarizmi's *Astronomical Tables* and *Liber Ysagogarum Alchorismi* around 1126. In 1142 he translated Euclid's *Elements* from Arabic, opening the door to Greek geometry and mathematics. He also attempted to put together much of the new knowledge of natural philosophy in *Questiones Naturales*, written in 1111. Stephen of Antioch (fl. 1120) translated Haly Abbas's *Liber Regalis*, a medical encyclopedia, in 1127. Robert of Chester (fl. 1140) followed Adelard's mathematical work with a translation of al-Khwarizmi's *Algebra* in 1145. Eugenius of Palermo (fl. 1150) translated Ptolemy's *Optics* in 1154, and Henricus Aristippus (fl. 1150) finished Aristotle's *Meteorologica* in 1156. Galen was translated into Latin by Burgundio of Pisa around 1180, introducing another aspect of natural philosophy through medicine.

It was an exciting time for scholars as new knowledge was uncovered one manuscript at a time from the treasure trove of Arabic sources. One of the most important conduits for Greek and Arabic natural philosophy was the school of translation established by Archbishop Raymond at Toledo after its fall to Christian forces. Toledo was an ideal location, since it had long been a meeting place for Christian, Jewish, and Islamic scholars. It was there that Gerard of Cremona (1114–87) discovered the astronomical work of Ptolemy and translated the *Almagest* in 1175, placing the best of Greek astronomical knowledge in European hands. Gerard translated over 80 other works during his lifetime, including the works of al-Kindi, Thabit ibn Qurra, al-Razi, al-Farabi, Avicenna, Hippocrates, Aristotle, Euclid, Archimedes, and Alexander of Aphrodisias.

Natural philosophy represented only a small portion of the rediscovered texts, but it gave the intellectual class of Europe a greater taste for the ancients, whose work they were eager to adopt and adapt. Cicero and Seneca were popular, while Aristotle's system of logic was significant in a wide range of applications. The natural philosophers of the twelfth century privileged Plato's *Timaeus* over Aristotle's works, because Plato's idealism accorded well with Christian theology. Among Jewish scholars of this period Moses Maimonides (1135–1204) was the best known; his *Dalalat al-Hairin* (*Guide to the Perplexed*) attempted to place Jewish philosophy on a firm Aristotelian foundation. Written in Arabic (Maimonides was a physician at Saladin's court), it was translated into Hebrew and later into Latin.

Not all medieval scholars restricted their research to intellectual material and the rediscovery of the works of the ancient philosophers. There was enormous interest in the promise of the manipulation of nature offered by the alchemical texts. When Robert of Chester translated *Liber de Compositione Alchemie* (*Book of the Composition of Alchemy*) in 1144, he introduced alchemy to Europe. A compendium of Arabic chemistry, it was followed by a flurry of research as people hunted for more detailed work by Jābir (Geber) and al-Razi (Rhazes).

Early in the thirteenth century there was another burst of university founding that took advantage of the new knowledge and the growing market for education. The University of Padua was founded in 1222 and became a leading medical school. The University of Naples followed in 1224, with the University of Toulouse close behind in 1229. Starting in 1231 Cambridge became Oxford's chief rival. The University of Rome was founded in 1244, and the Sorbonne University in 1253.

Christian Theology vs. Aristotle's Natural Philosophy

The universities soon established themselves as *the* site of intellectual activity in Europe. While autodidacts (those who were self-taught) and those from earlier cathedral schools might once have claimed equal footing as scholars, by the end of the thirteenth century the Professor of Theology had much higher status. In this way the universities became both the protectors and creators of knowledge. However, they were essentially conservative institutions, so once something was made required reading, it became an unchallengeable authority. At the same time the universities stood in a complex relationship to the larger structure of the Catholic Church. They were seldom under the complete control of any one bishop, and thus they provided space that was sanctioned by the Church and yet not controlled by it. This allowed the debate about the primacy of faith or reason to be played out within their walls and cities. While several scholars were imprisoned for their impious views, the fact that these debates could take place at all speaks to the power and independence of these institutions.

Christian theologians were not universally pleased that the work of Arabic and Greek philosophers was being introduced to the Latin West. Aristotle was particularly subject to theological objections since he contradicted the Bible on many issues of natural philosophy, such as the infinite life of the universe, and as a pagan he offered an implicit challenge to Christian authority. Because of Aristotle's popularity among students, authorities at the University of Paris grew concerned over the effect of pagan philosophy on the future theologians and secular leaders being trained there; so, in 1210 they banned the reading and teaching of his works on natural philosophy. This was also a battle over authority, since the conservative Faculty of Theology effectively imposed the banning of Aristotle on the more progressive Faculty of Arts. The ban was renewed in 1215 by Robert de Courçon, a papal legate, and again by Pope Gregory IX in 1231. However, the general interest in Aristotle prompted Gregory to establish a commission to review Aristotle's work and clean up any theologically problematic elements.

Ironically, the banning of Aristotle's natural philosophy actually promoted the study of it, making it a kind of philosophical forbidden fruit. The ban applied only to the University of Paris, so other universities were free to offer Aristotelian instruction, and this was used as a selling feature to attract students. Further, the ban covered only natural philosophy, so Aristotle's work on logic, despite being intimately bound to the system of natural philosophy, was still available for study. Demand for Aristotle continued to grow, and the supply of texts and scholars also multiplied. Finally, in 1255 pressure to learn Aristotle and the wide availability of texts led the Faculty of Arts at Paris to pass new statutes that made instruction in Aristotle not just acceptable but a mandatory element of an arts education. Aristotle's works had gone from being outlawed to required knowledge in just 45 years.

The work of Aristotle became so fundamentally important to the intellectual life of the Latin West that he was referred to simply as "the Philosopher." Although translation efforts continued, the difficulty of his arguments and the often fragmentary nature of the available texts led to a heavy dependence on Arabic commentators.

In the early period of reintroduction the most popular commentator was ibn Sina (Avicenna). By the middle of the thirteenth century Rushd (Averroes) had become the chief commentator used by Latin scholars. Like Aristotle, Rushd was so important that he was referred to as "the Commentator."

This exalted treatment of Aristotle and his commentators gives us a false picture of medieval scholarship if it is taken to suggest a slavish or doctrinaire dedication to the Greek material. Historians for many years argued that medieval scholarship was largely derivative and thus an uninteresting but necessary path to later work that challenged Greek ideas. More recently, historians have realized that, while dedication to the texts was an important element of medieval scholarship, from the earliest times there was a constant debate about every aspect of Greek natural philosophy. One of the major concerns to plague medieval scholars was that the Greeks were not Christian, so every aspect of ancient philosophy had to be debated in the light of Christian orthodoxy. Since the majority of Latin scholars were members of the clergy, the pagan origin of Greek thought was seen by some as reason enough to reject it; this was part of the motivation for the banning of the study of Aristotle.

A more moderate group as typified by Pope Gregory IX was prepared to include aspects of Greek thought as long as they were not overtly theological or directly contradicted biblical authority. Indeed, one of the first challenges for medieval philosophers was to find a way by which Greek natural philosophy could coexist with revealed religion. The latter was necessary for salvation, but the former offered a path to understanding God's creation as well as a wealth of practical knowledge. Among those who made a formal attempt to align Aristotelian philosophy with Christian theology was Robert Grosseteste (c. 1168–1263). Grosseteste was the first chancellor of Oxford University and a man of enormous intellectual breadth. He worked on Aristotle's logic in the *Posterior Analytics* and on his physics and mechanics from *Physics, Metaphysics*, and *Meteorology*. Grosseteste reconciled Aristotelian ideas with biblical thought in commentaries on logic and natural philosophy. For example, he argued that while creation by God took precedence over the cosmology used by Aristotle, it did not follow that Aristotle was wrong about the composition of matter in the universe.

Grosseteste was also deeply interested in optics, working with Euclid's *Optica* and *Catoptica* as well as al-Kindi's *De aspectibus*. This fascination with light came in part from a belief that light in the material world was analogous to the spiritual light by which the mind gained certain knowledge about true forms or the essence of things. Light was the fundamental corporeal substance, and so the study of optics was the fundamental study in natural philosophy. Since understanding optics required mathematics, Grosseteste linked mathematics, natural philosophy, and religion together. His teaching, particularly to members of the Franciscan Order, led many scholars to the study of mathematics and natural philosophy.

Following Grosseteste was the great medieval thinker Albertus Magnus (c. 1206–80). Albertus held one of the two Dominican professorships at the University of Paris and was keenly interested in finding a place for Greek philosophy within the context of the Church and in challenging the intellectual place of the Franciscans. He wrote extensively on philosophy and theology and is remembered for many works on natural topics, ranging from geology to falconry and the powers of plants and magical beasts. Albertus was an energetic scholar who wrote commentaries on all available Aristotelian texts. Because of the range of his work, he became known as "Doctor Universalis," and he was not afraid to amend or correct "the Philosopher" on either natural or philosophical issues. Albertus did not propose a new orthodoxy based on Greek philosophy, but he argued that a corrected natural philosophy had great utility and could be exploited by the existing orthodoxy. As such, he expected the intellect to glorify the creation of God and the utility of natural philosophy to aid in making Christianity supreme.

Magic and Philosophy

Albertus Magnus was also the supposed author of one of the most popular medieval texts, *Liber Aggregationis*, or, by its English title, the *Book of Secrets*. The text was written by an unknown author or authors, perhaps even students of Albertus, and attributed to him. It is a compendium of treatises on "herbs, stones, and certain beasts," and from a modern perspective it seems to be mostly about magic, astrology, and mythical beasts such as the cockatrice and the griffin. Yet the work tries to set the world into a loosely Aristotelian framework, and there are aspects of both Pliny's encyclopedic descriptions of the world and the material inquiries of the Islamic alchemists in it as well. While most serious scholars of the age disdained this kind of mysticism, such compendiums were enormously popular. A large section is set in a kind of problem/solution format, offering formulas and methods of procedure to deal with specific problems, such as this defence against drunkenness:

If thou wilt have good understanding of things that may be felt, and thou may not be made drunken.

Take the stone which is called Amethystus, and it is of purple colour, and the best is found in India. And it is good against drunkenness, and giveth good understanding in things that may be understood.[1]

The *Book of Secrets* is a book of medieval magic and contains a powerful, if ill-defined, link between magic and natural philosophy. At the simplest level both were studies of an unknown world, and both offered the possibility of controlling the unknown through naming and describing it. Yet the magic of the *Book of Secrets* is instrumental rather than spiritual, and this distinction was important for practitioners interested in the unseen forces and powers of nature. The *Book of Secrets* carefully avoids the issue of witchcraft, supernatural powers of either good or ill, or calling on the powers of supernatural beings. As fantastic as were the properties of the items listed and described, they existed in the objects themselves, they were hidden except to the knowledgeable, and they were natural.

One of the most notable elements of the text is that it uses the terms *experimentari* and *experiri*, referring to experiments rather than just the experience of nature. While it is difficult to assess if or how much these descriptions and recipes were believed, it was certainly the case that many people took them seriously enough to try them. Even if Albertus was not the author of the *Book of Secrets*, he did favour a form of Aristotelian natural philosophy that was shaped by Arabic tradition and that included a more hands-on approach to the study of nature. This was a departure from the approach of earlier Latin scholars who were more interested in knowing what could be true than in knowing what something looked like, or how it might react when mixed with other things.

Roger Bacon and Thomas Aquinas: The Practical and Intellectual Uses of Aristotle's Natural Philosophy

The path of natural philosophy split after Grosseteste and Albertus Magnus. Those more attracted to the investigative side of the subject, such as Roger Bacon (c. 1214–94), began to copy the practical approach of many of the Arabic sources.

1. Albertus Magnus, *The Book of Secrets of Albertus Magnus*, ed. Michael R. Best and Frank H. Brightman (Oxford: Oxford University Press, 1973) 33–34.

This group included the growing number of alchemists and astrologers. Those more interested in philosophy and an adherence to the Greek intellectual tradition tended to see the subject in terms of its ability to train the mind and provide ways of gaining certain knowledge. This stream led to Thomas Aquinas (1225–74) and the scholastics. A third stream can be seen in the spread of primarily practical skills among the engineers, masons, smiths, navigators, and healers of the Middle Ages. This group has been overshadowed by the others because they were rarely part of the intellectual class and left few written records; yet, it is clear that everything from the construction of the cathedrals to the practice of midwifery was affected by natural philosophy as it filtered through European society.

Roger Bacon is a perfect example of the spirit of enquiry in the Middle Ages. He studied at both Oxford and Paris and later joined the Franciscan Order. He favoured the utility of natural philosophy, especially that found in Aristotle's more practical works, and argued that the comprehension of nature would aid Christianity. He wrote on optics, speculated about the design of underwater and flying vehicles, and supported the idea of experiment as a method of discovering things about

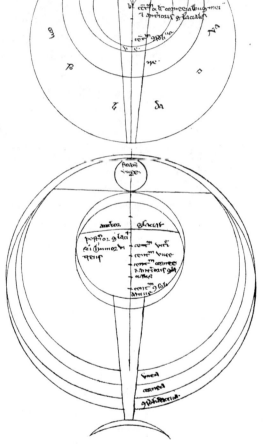

3.2 ROGER BACON'S OPTICS

Bacon's version of Alhazen's optics of the eye. In the top circle, light enters the eye from the top and passes through the vitreous humour. The bottom circle is a detail of the interior of the eye.

Roger Bacon's *Optics Diagram of the eye*, from the work of Roger Bacon/Universal History Archive/UTG/Bridgeman Images.

nature. He was the first European to mention gunpowder, but it is uncertain whether this was an independent discovery or learned from Eastern sources (having been discovered around the ninth century in China) and recreated by him. The text in which he mentioned gunpowder was his *Opus Majus*, written around 1267. The book was not published until 1733, and it is unclear how widely the manuscript circulated in his day. His speculations and defence of natural philosophy were not well received by the Franciscan leadership, but he persisted, believing that he had a duty to pursue his work. Eventually he was reprimanded, put under surveillance by his Order, and finally imprisoned for heresy in 1277.

The greatest figure of the intellectual stream was Thomas Aquinas. He had been a student of Albertus Magnus, following his teacher's lead in clarifying the interaction between theology and philosophy. For Aquinas, faith and the authority of God were primary, but in those areas not determined by revelation God had granted humankind the tools to understand nature. Thus, there could be no true conflict between religion and philosophy, since God had given us both. Any apparent contradictions disappeared when proper theology and proper philosophy were applied. Aquinas followed the philosophic path begun by Rushd (Averroes) and, in a sense, saved Aristotle by compartmentalizing his work. In one box he put Aristotle's system for gaining certain (or true) knowledge and the method of testing knowledge based on logic. If results were arrived at through the proper methods, the product of philosophy could not contradict revelation. He placed Aristotle's observations about the world in another box. These contained some erroneous material, but the big picture—such as the perfection of the heavens—was correct, and, as such, many of the observations were worth the effort of cleaning up or Christianizing. In the last box were Aristotle's ideas about theology, politics, and social structure. These, along with errors by other pagans, were disregarded as being heretical, false, or superseded.

The discussion of Aristotelian philosophy indicates in part how important Greek philosophy had become for the intellectual community of Latin Europe. Aquinas's work was situated within a serious scholarly debate about the place of philosophy (Aristotle's work in particular) in the intellectual arena, but it was also written to counter a number of specific challenges to orthodoxy. One of his chief targets was Siger of Brabant (c. 1240–84) who held a strongly Aristotelian view of the world and attempted to teach philosophy without the constraint of theology. In response, Aquinas wrote *On the Unity of the Intellect, against the Averroists*, which, while specifically attacking Siger's position, more generally

PLATE 1: CELESTIAL
CLOCK OF GIOVANNI
DE DONDI OF PADUA

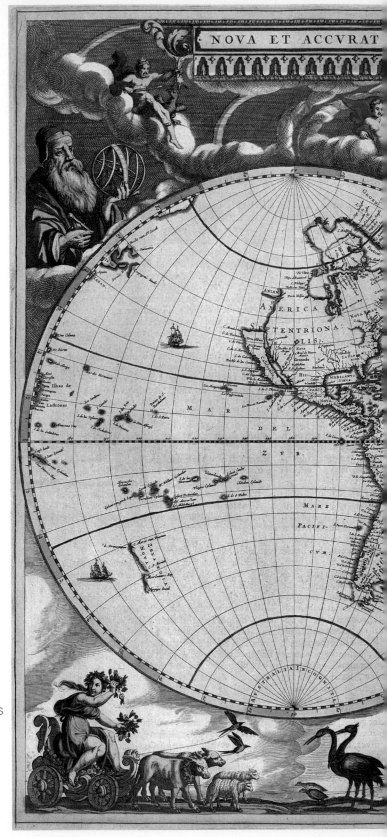

PLATE 2: JOHANNES BLAEU'S
WORLD MAP, C. 1664

© Corbis

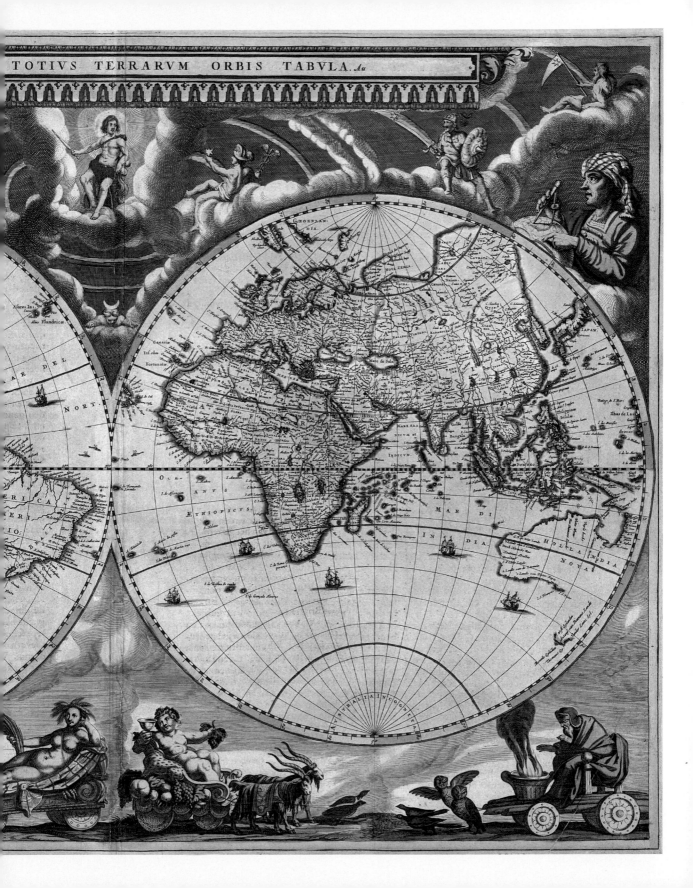

TOTIVS TERRARVM ORBIS TABVLA. *Au*

PLATE 3: HOLBEIN'S
THE AMBASSADORS (1533)

National Gallery, London,
UK / Bridgeman Images.

PLATE 4: "AN EXPERIMENT
ON A BIRD IN AN AIR PUMP,"
JOSEPH WRIGHT OF DERBY
(1768)

National Gallery, London,
UK / Bridgeman Images.

argued that philosophy was dependent on theology and should not stand alone. Aquinas won, and Thomistic natural philosophy became the orthodoxy of the European scholarly world.

Aquinas's writing and reasoning were dense, even compared to contemporary medieval scholars, and this in turn made his work the focus of much study. Consider this short passage from the Introduction to his work *On Being and Essence*:

> Moreover, as we ought to take knowledge of what is simple from what is complex, and come to what is prior from what is posterior, so learning is helped by beginning with what is easier. Hence we should proceed from the signification of being to the signification of essence.[2]

While it seems reasonable to move from what is easy to understand to what is complex, Aquinas's idea of what was easy and complex has provided scholars with 700 years of debate.

Scholasticism

By the beginning of the fourteenth century the study of Aristotle in the Thomistic tradition was in complete ascendancy. While there were still evangelical members of the Church who questioned any worldly study as a distraction from faith, Aristotelianism had flowed into every aspect of intellectual life and had taken up a position alongside the Church fathers as a source of authority. The intersection of Aristotelian methodology and medieval interests including theology and certain aspects of Platonic philosophy developed into a form of philosophy known as scholasticism. The scholastics were closely associated with the universities and the more intellectually inclined religious orders such as the Dominicans.

Scholasticism represents the strongest vein of intellectualism in the Latin Church and can be traced back to Augustine in the fourth century. In the early medieval period it owed more to Plato and his idealism than to Aristotle, who was

2. Thomas Aquinas, "On Being and Essence," in *Philosophy in the Middle Ages*, ed. Arthur Hyman and James J. Walsh (Indianapolis: Hackett, 1984) 508.

little known except for his logic. The basic method for the scholastics was the dialectic, so that questions were posed in such a way as to establish two contradictory positions. The idea of resolving a question by presenting contradictory initial positions was well known to the Greeks and makes up the basis of the dialogue form used by Socrates, but the medieval scholars took this method to new levels of intricacy. This began with a formalized organization of the argument into thesis, objections, and solutions. Peter Abelard's (1079–1142) *Sic et non* (*Yes and No*) was one of the seminal texts for this method. Thomas Aquinas used it in his reconciliation of Aristotle with Christianity.

Herein lies the historical problem of the scholastics, since their dedication to Aristotle became so strong that, over time, their system was transformed from a method for understanding the world into an axiomatic statement about the world. The scholastics were rationalists at heart in that they argued reason was required to understand the universe, but they created a system that relied on a set of authorities that were then largely placed beyond question. This did not mean that debate ended, and in fact it remained one of the fundamental skills for scholars. Universities took up the *dissertatio* as the method to achieve higher degrees. This skill extends to the modern day with the dissertation and defence system used to obtain a PhD, a doctorate in philosophy. The system supposes that the thesis is an argument made by a student who publicly defends it against questions posed by scholars knowledgeable in the field; our continued use of the method created by the medieval scholars indicates how robust an educational system it is. There was a limit, however, to the amount of new insight that could be gained by perpetually debating the same issues. Thus, the debates were less concerned with reaching a conclusion and were seen more as a tool to train novices to understand the established answer or to bash opponents back into line.

Medieval Alchemy

While Aristotle was undergoing theological and philosophical revision on the road to orthodoxy, another conduit for investigations of the natural world was not being subjected to the same process of legitimization, because it was not a part of the school system. More than medicine, astronomy, mathematics, or philosophy, alchemy brought an interest in natural philosophy to a wide audience that included princes, physicians, teachers, monarchs, religious leaders, craftspeople, and commoners. If only by sheer numbers, alchemists were the most common proponents of the study of nature. Because the Islamic alchemists based their work on a version of Aristotelian matter theory, the works of Rhazes and Geber were the greatest conduit of Greek natural philosophy as a practical art into the Latin West. This had positive and negative effects on natural philosophy. On the positive side, it expanded the study of the material world and was instrumental in introducing skills and the concept of experiments. The negative aspects were, first, the degree of charlatanism that came to be associated with it, and, second, the alchemists' secrecy and even paranoia, which was contrary to the concept of public knowledge so characteristic of natural philosophy.

The medieval charlatans were many, and they played on the greed and gullibility of both the high and low born. The basic con was simple. The charlatan claimed to have discovered the process for creating the Philosopher's Stone, thus persuading a rich benefactor to support the actual production of gold from base metals. During the production, the alchemist was housed, fed, and clothed by the patron and might even be given a stipend to cover other living expenses. Also, costly and exotic materials were needed. Since alchemical knowledge was arcane and secret, who among the victims could say that the expensive white powder was not a rare ingredient made from the feathers of a phoenix and imported from Cathay? In addition to the money made indirectly, the alchemist often required quantities of gold as a seed for the transformation of undifferentiated prime matter into the precious metal.

Charlatan alchemists found ready victims. The medieval world was full of fantastic beasts, evil spirits, and magicians, so alchemy fit with the belief in the existence of supernatural forces. In addition, transmutation of matter was preached as doctrine by the Church. In transubstantiation the Eucharist bread and wine were transformed into the body and blood of Christ, while many biblical stories hinged on transformation of matter in some way, such as Lot's wife turning from flesh to salt, Eve being created from Adam's rib, or Christ changing water into wine. While the Church outlawed witchcraft and regarded magic as dangerous and evil, it was through the Church that alchemy came to Europe, was translated and transcribed in the *scriptoria*, was studied by popes and cardinals, and was practised by monks.

What complicates the story of the alchemists was that the "true" alchemists (those who were not simply con artists) also needed patrons and funds to carry out their work. If that meant occasionally improving results to placate patrons, that was the price of research. There were also the contradictory pressures on the alchemists to keep their processes secret (for personal and financial reasons) and the necessity of making their work public in order to attract patrons. This continues to be a problem even today, when the pressure to produce results has occasionally led scientists to fabricate or adulterate results (or at least produce conclusions far beyond their evidence) in order to secure funding for their experiments.[3]

Arnold of Villanova (c. 1235–c. 1311) is a good example of a true medieval alchemist. Famous as a physician, he was also an astrologer and alchemist. He wrote a treatise on transmutation called *The Treasure of Treasures, Rosary of the Philosophers and Greatest Secret of All Secrets*, in which he claimed to have found the secret of matter known to Plato, Aristotle, and Pythagoras. He told his readers that he would hold nothing back, but that they must read other books to understand the hidden reasoning behind his work. Transmutation could be achieved through a kind of purification of metal that would leave behind only the noble elements of silver and gold. This was to be accomplished by an *aqua vitae* (water of life) made from mercury, which in turn was used to produce an elixir that could convert a thousand times its weight in base metal into gold or silver (depending on the elixir). The process was described in terms of the life of Christ, covering conception, birth, crucifixion, and resurrection. While most of the material was theoretical, there was enough practical direction (and evidence of actual work) to encourage readers to attempt to replicate Arnold's work.

Experiment and Explanation

While alchemy was one way of investigating material that was not dominated by ancient philosophy, medieval scholars were themselves quietly examining nature and finding Aristotelian observations wanting. They were not as slavishly devoted to the Aristotelian texts as they may appear to be, given the effect of scholasticism. By using Aristotelian methodology, medieval scholars challenged what was true knowledge without risking an attack on authority, especially if they concentrated on the observational material in the compartmentalized Aristotle. In a typical approach, the natural philosopher would begin by praising Aristotle and then

3. A modern example of this sort of wishful thinking can be seen in the attempt by Pons and Fleischmann to create cold fusion.

proceed either to explore an area that he had not covered or to demonstrate a new idea in the guise of a moderate correction to his impeccable system.

This can be seen in the work of people such as Robert Grosseteste and Theodoric of Freiberg (c. 1250–1310) who both worked on optics and the rainbow. Aristotle argued that the rainbow was the result of sunlight reflecting off water droplets in clouds that acted like tiny mirrors. Arab work on optics with its more practical aspects showed in contrast that the rainbow was created by refraction. Grosseteste began his examination as follows:

> Investigation of the rainbow is the concern of both the student of perspective and the physicist. It is for the physicist to know the fact and for the student of perspective to know the explanation. For this reason Aristotle, in his book *Meteorology*, has not revealed the explanation, which concerns the student of perspective; but he has condensed the facts of the rainbow, which are the concern of the physicist, into a short discourse. Therefore, in the present treatise we have undertaken to provide the explanation, which concerns the student of perspective, in proportion to our limited capability and the available time.[4]

Thus, Grosseteste argued that he was not demonstrating that Aristotle was wrong about the rainbow; rather, he was merely filling in that part of the investigation that Aristotle did not cover. This was a common ploy for scholastic natural philosophers, allowing them to maintain their allegiance to the Philosopher while they presented original work without fear of being accused of hubris for placing their work above his.

Theodoric praised Aristotle and then tossed aside his theory to present his own, one based on refraction and reflection. This was likely based on material he learned from Alhazen's *Book of Optics*. He offered a method of testing the behaviour of light that falls on a raindrop by obtaining a glass globe, filling it with water, and shining a light on it. (See figure 3.3.) While Theodoric's work was not the first example of experiment in the Latin West, it is often pointed to as a precursor to experimentalism, particularly because his results are essentially those we find today. It is a good example of the kind of intellectual bridge between Aristotelian philosophy and the move to test observations that would transform the study of nature.

....

4. Robert Grosseteste, *On the Rainbow*, "Robert Grosseteste and the Revival of Optics in the West," in *A Source Book in Medieval Science*, ed. David Lindberg and Edward Grant (Boston: Harvard University Press, 1974) 388–89.

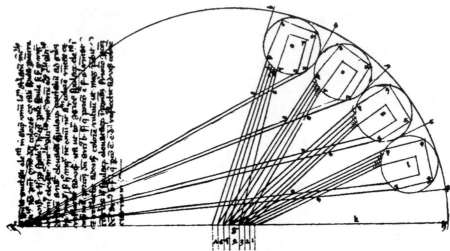

3.3 THEODORIC'S RAINBOW FROM *DE IRIDE* (C. 1304) The small circles at the right represent raindrops that reflect and refract the light entering from the left and seen at the middle.

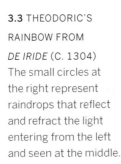

In Aristotle, the truth about nature is to be found in the intellectual construct that results from the application of logic to observation. In other words, we know the truth because of our ability to apply a system of classification and explanation to sense perception. Theodoric does not deny the Aristotelian system, but he pushes against the Aristotelian location of sense perception. The unaided eye cannot discern the correct sense perception, so the creation of the rainbow must be modelled in such a way that the event can be made clear to the senses. The glass globe is not a raindrop, but Theodoric makes the implicit assumption that it must be analogous to a raindrop and thus must represent the physical condition of the raindrop. The truth about the rainbow no longer lies solely in the observer (the senses and the intellect) but must also reside in the apparatus that replicates the physical conditions.

While we have come to accept the kind of reasoning behind Theodoric's work, it was not self-evident that certain knowledge could be gained by such a method. One of the principal problems of reasoning from observation was the impossibility of certainty by induction. By definition, sense perception relies on induction: an observer noticing only white swans might reasonably go from a series of particular observations to the general conclusion that swans could only be white. Since observation cannot limit the possibility of a black swan, nor can the observer know that all possible swans have been seen (since that would include both past and future swans), the best that can be said is that all observed swans are white.

Likewise, in the Middle Ages, the argument Aristotle had made against experiments was still taken seriously. That is, forcing nature to perform unnaturally (in an experiment) does not give one insight into its natural behaviour.

Ockham's Razor

Skepticism about the possibility of certain knowledge as formulated in the Aristotelian/scholastic system was not uncommon. The primary source of attack came from mystically inclined theologians who objected to rationalism and logic altogether, but there were philosophic challenges as well. The most forceful skeptic was William of Ockham (1285–c. 1349), who attacked the Aristotelian categories of relation and substance, thereby undermining both physics and metaphysics. Ockham argued that relations were created in the mind of the observer and did not represent any underlying order in the universe. Thus, Aristotle's four spheres of elements existed only in the mind, collapsing the whole edifice of Aristotelian explanation. Ockham also challenged Aristotelian teleology, arguing that it was impossible to prove by experience or logic based on first principles that there was a final cause for any particular thing. Part of Ockham's defence of this philosophy was based on the Law of Parsimony, more commonly called "Ockham's Razor." He argued that "plurality should not be posited without necessity."[5] In more direct language this meant that an explanation of some problem would not be made better by adding arguments to it. As a philosophical device, it also suggested that, when faced with more than one explanation for a phenomenon, it was wise to choose the simplest. This idea was not originally Ockham's (versions of the idea of philosophical parsimony can be found in the work of Maimonides and even Aristotle), but it was one of his guiding principles. Much of Aristotle's elaborate system seemed to Ockham to be unnecessary or unprovable.

In addition to challenging scholasticism Ockham also challenged the hierarchy of the Church. He believed that revelation was the only source of true knowledge, and this belief set him at odds with the policy of papal authority. Although willing to accept the supremacy of the Church on spiritual matters, he objected to the extension of papal authority to secular issues such as the subordination of

5. William of Ockham, *Summa totius logicae* (c. 1324), in *Philosophy in the Middle Ages: The Christian, Islamic, and Jewish Traditions*, third edition, ed. Arthur Hyman, James J. Walsh, and Thomas Williams (Indianapolis: Hackett, 2010), p. 624.

monarchs to the Church's temporal authority. For his loud and public objections, he was excommunicated on June 6, 1328. At that time he was under the protection and patronage of Holy Roman Emperor Louis IV, so was protected from the wrath of the pope.

The Okhamites were few in number, in part because the position was dangerous politically, but they had a wide impact. Their philosophy was labelled "nominalism" because it denied the actual existence of abstract entities or universals. By extension, the natural world could only be described in "contingent" terms. Something that is contingent might be true or might equally be false. Consider the statement "all swans are white." This conclusion could be reached by observation and held to be universally true, but such a conclusion was proven false when the *Cygnus atratus* or Australian black swan was discovered. The matter is contingent on factors external to the proposition. If universals did not exist and nature was contingent, then the only way to discover anything about the natural world was through observation, and all general statements (such as classification) were potentially subject to revision based on further observation. This philosophy was part of a trend by some philosophers away from the study of the traditional realm of metaphysics and toward the study of experience. Moreover, the Ockhamite position suggested the independence of philosophy from theology. Although it was not the only group to do this, the nominalists opposed the majority of medieval thinkers who accepted Aquinas's position that philosophy was subordinate to theology. This was another form of the separation of the natural from the supernatural, which was a necessary step if there were to be an independent study of natural philosophy.

The Black Death and the End of the Middle Ages

Ockham's death occurred just before the greatest natural disaster of the Middle Ages: the plague or Black Death. The plague started in China in the 1330s and was carried by traders to the Black Sea, where Italian merchants, sailors, and shipboard rats were infected and passed it on to Europe in 1347. The disease was horrific, spreading through air, by touch, or by flea bite. People often died within hours of exposure. It was called the Black Death because it caused buboes (hence bubonic plague), or swellings filled with dark blood, to form on the body, especially near the lymph nodes in the groin, armpits, and throat. The Italian author Boccaccio wrote that victims "... ate lunch with their friends and dinner with their ancestors in paradise."[6] Many historians place the death toll at 25 million in five years, or one-third of Europe's population, but the figure may have been as high as 50 per cent. Many towns and villages, where the proximity of people led to a rapid spread of the disease, were totally depopulated. The effect of the plague was made worse because the people of Europe had experienced a series of bad harvests before it arrived, and malnutrition and starvation had already weakened the population.

The appearance of the plague coincided with the Hundred Years' War (1337–1453) that pitted France against England. England lost most of its continental lands, but the prolonged conflict wiped out a significant portion of France's nobility. The Great Schism (1378–1417) also followed on the heels of the Black Death and led to the central authority of the Church splitting, as competing popes in Rome and Avignon attempted to rule at the same time. All the death and destruction of the era encouraged a swing toward a more conservative theology and promoted a resurgence in mystical Christianity. The Black Death certainly seemed like a biblical curse, and no earthly action had any effect on it. Physicians often blamed disease on bad astrological events, and the medical faculty at the University of Paris concluded that the plague was the result of a conjunction of Jupiter, Saturn, and Mars that corrupted the air. The "calamitous 14th century," as historian Barbara Tuchman called the era,[7] marked the beginning of the end of medieval Europe. Although it took almost 400 years for the social structure of the Middle Ages to fade completely from Western European society, the new path was opened not by philosophers, social reformers, merchants, monarchs, or popes, but by the misery of disease.

In the plague years less original work was done in natural philosophy, since most theologians and scholars who did survive were more concerned about death and salvation than about the structure of nature. Nicolas Oresme (c. 1323–82), the Bishop of Lisieux, was one of the few who continued to work on natural philosophy. Oresme's work on mathematics was a precursor to analytical geometry as he tried to represent velocity geometrically. In *Le Livre du Ciel et Monde*, a commentary on Aristotle's *De Cœlo*, he presented the most comprehensive examination of the possible motion of the Earth to date, concluding that the evidence supported the geocentric model of Ptolemy. Oresme wrote *Ciel et Monde* and translated a number of Aristotle's works into French at the command of Charles v, and as such his work marked a shift in attitude toward the use of vernacular rather than Latin.

....

6. Boccaccio, "Introduction," *Decameron* (1351) Day I.

7. Barbara Tuchman, *The Distant Mirror: The Calamitous 14th Century* (New York: Alfred A. Knopf, 1978).

The greatest effect of the plague on natural philosophy was indirect. The death of so many people meant that when the plague years passed, the land was vastly underpopulated. For those who survived, life held many more possibilities than it had before. The survivors inherited the property of the victims, and many people grew suddenly rich as they gained the inheritance not only of their immediate family but often of distant relatives as well. Good land was plentiful, but the people to work it were scarce, so peasants got better deals from landowners and could afford to buy more luxury items. It was also easier for peasants to leave the land and enter into trades and mercantile activities. Cities, countries, and the wealthier nobles often had to compete to attract artisans and even peasants to their regions. The booming economy made rich those people who could supply the demand for luxury items such as silk and fine cloth, spices, ivory, perfume, glassware, jewellery, and a huge list of manufactured items from footwear to armour to mechanical toys. In the leading centres of commerce, particularly the Italian city-states of Genoa and Venice, this new money paid for merchant and naval fleets, public works, an explosion in patronage of the arts, and education. The people outside the trade centres saw their gold and silver flowing out of their regions and making others rich, as Italian merchants and their Arabic trade partners controlled the flow of the most expensive luxury items that came from the Far East. The Spanish, Portuguese, English, and the Italians themselves began to consider ways to get around the middlemen and trade with China directly.

To do that the Europeans needed a host of tools: better astronomy for navigation, improved cartography and geography, new and better instruments, and better mathematics to make these possible. Key to the new trade initiatives were new ships that could sail the Atlantic, so better naval engineering was required. But what they needed more than the tools were the people to devise them, build them, and take them out and use them. Natural philosophy was a key component to this drive, and, combined with Johannes Gutenberg's nifty invention of the printing press, Europe had all the elements necessary for an explosion of intellectual, economic, and cultural activity.

Conclusion

From the fall of Rome European rulers, Church leaders, and intellectuals had laboured to create a stable, hierarchical society. In intellectual terms, they began by creating a need for Greek philosophy, integrating it into their educational and theological world view. By 1300 Europe was growing slowly, and society was well ordered, carefully regulated, and somewhat inward-looking. Natural philosophy was studied by a small intellectual group primarily at the universities, while the alchemists, physicians, and artisans worked away at practical problems. For most thinkers, the demarcation had been established between philosophical and revealed knowledge. As Aquinas had shown, these two knowledge systems were not in conflict, since they dealt with exclusive areas of knowledge, with theology as the superior study and philosophy in a useful but supporting role. In other words, the Latin scholars had faced the same issue as the Greeks and had determined the separation between the supernatural world of revealed religion and the natural, rationally understood world of nature. The tension inherent in this separation was a very productive one, allowing some of the finest thinkers to create the impressive intellectual system of scholasticism. At the same time those interested in the application of this knowledge to practical ends lived more in the world than the academic scholastics. By 1450 their time had come. European society had been shaken by its encounters with the four horsemen of the Apocalypse, but in the aftermath a new sense of prosperity and freedom emerged. There was adventure in the air.

Essay Questions

1. How and why did Charlemagne support education?

2. How did Christian scholars overcome the inherent problems with Aristotelian philosophy and why did they do so?

3. How did alchemists contribute to the spread of natural philosophy?

4. In what ways did the Crusades transform the study of natural philosophy in Europe?

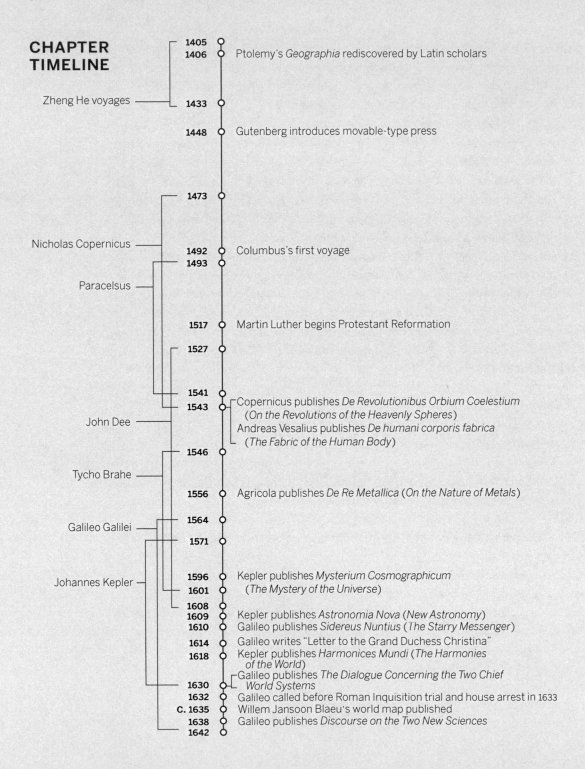

CHAPTER TIMELINE

1405
1406 — Ptolemy's *Geographia* rediscovered by Latin scholars

Zheng He voyages — **1433**

1448 — Gutenberg introduces movable-type press

1473

Nicholas Copernicus — **1492** — Columbus's first voyage
1493

Paracelsus

1517 — Martin Luther begins Protestant Reformation

1527

1541 — Copernicus publishes *De Revolutionibus Orbium Coelestium*
1543 — (*On the Revolutions of the Heavenly Spheres*)

John Dee — Andreas Vesalius publishes *De humani corporis fabrica*
(*The Fabric of the Human Body*)

1546

Tycho Brahe

1556 — Agricola publishes *De Re Metallica* (*On the Nature of Metals*)

Galileo Galilei — **1564**

1571

Johannes Kepler — **1596** — Kepler publishes *Mysterium Cosmographicum*
1601 — (*The Mystery of the Universe*)

1608
1609 — Kepler publishes *Astronomia Nova* (*New Astronomy*)
1610 — Galileo publishes *Sidereus Nuntius* (*The Starry Messenger*)

1614 — Galileo writes "Letter to the Grand Duchess Christina"
1618 — Kepler publishes *Harmonices Mundi* (*The Harmonies of the World*)

Galileo publishes *The Dialogue Concerning the Two Chief World Systems*
1630
1632 — Galileo called before Roman Inquisition trial and house arrest in 1633
C. 1635 — Willem Jansoon Blaeu's world map published
1638 — Galileo publishes *Discourse on the Two New Sciences*
1642

SCIENCE IN THE RENAISSANCE: THE COURTLY PHILOSOPHERS

4

The intellectual life of Europe expanded in the fifteenth and sixteenth centuries. Natural philosophers found new texts, new lands, new interpretations, and new career paths. The European Renaissance, meaning "rebirth," began with a renewed interest in the discovery of classical texts. This intellectual voyaging was matched by a greater confidence and spirit of adventure that led to contacts with newly discovered peoples and places. Europeans found that they were living in a world of expanding possibilities. They encountered people who had their own knowledge, particularly of navigation. While intellectuals first looked backward, to the glorious heritage of the ancients, they soon used ancient knowledge as a stepping stone to new information and ideas. At the same time the Catholic Church lost its professed monopoly on truth with the upheaval of the Reformation, while university scholastics found themselves under attack, no longer the sole controllers of philosophic knowledge. A window of opportunity was created, especially through patronage in the princely courts and merchant halls. Because of this, different things began to be valued. Rather than syllogistic logic and theological subtleties, princes wanted spectacle, power, and wealth. Therefore, natural philosophers who were practical (or claimed to be) were valued.

The Early Renaissance: Humanists and the Printing Press

As we have seen, Europeans had never completely lost touch with Greek knowledge and natural philosophy. They had studied Aristotle intensively during the Middle

Ages, to the point where his logic and larger intellectual system had become a foundational requirement for academic and theological discourse. However, there were large sections of the Greek and Roman corpus that had disappeared from view. Plato, especially, was largely unknown to European intellectuals, as were many other works of literature and philosophy. European scholars' eyes were opened by the rediscovery of, and engagement with, these great ancient thinkers. The men and women responsible for this rebirth were called humanists.

Beginning in the fourteenth century in Italy, scholars unaffiliated with the Church or the universities began to sell their services as teachers to the children of the rich and powerful in the Italian city-states. They taught humane letters, *studia humanitatis*, and stressed the *trivium* through the study of the great Latin writing of the past. Scholars such as Petrarch (1304–74), Leonardo Bruni (d. 1444), and Guarino da Verona (Guarino Guarini) (1374–1460) looked to the wonderful prose of Cicero and Seneca in order to understand how to be the good citizen and live the good life. Because these teachers changed the venue and purpose of education, both women and men had access to the new learning, and several women became well-known humanists. For example, Isotta Nogarola (1418–66) composed the "Dialogue on Adam and Eve," a debate as to whether Adam or Eve was more responsible for their banishment from Eden (seen as an early feminist discussion). Cecilia Gallerani (1473–1536) was a friend of Leonardo da Vinci and may have held the first "salon" or meeting of artists and intellectuals. She became famous as the "Lady with an Ermine" in the portrait by da Vinci. All these humanists were convinced that good words and thoughts made wise citizens, and they worked hard to find pure versions of ancient texts in order to achieve that wisdom.

This rediscovery of ancient wisdom and a reorientation to living a good life in this world rather than only working to achieve salvation in the next is often labelled the Renaissance. Although historians today hotly debate the use of the term, the period witnessed a flowering of intellectual and artistic activity that started in Italy during the fourteenth century and was emulated in other parts of Europe over the next 200 years. While humanists stressed the language-based studies of grammar, rhetoric, and logic, this changing intellectual world affected the study of nature as well. Scholars appeared who were willing and able to ask fundamental questions about the system of natural philosophy and to develop new methods of study. What started for these scholars as an intoxicating rediscovery of Greek natural philosophy ended in an almost complete abandonment of it. Over the period the scholars themselves underwent a radical change as they increasingly moved away from the Church and from theology as the foundation and reason for the study of nature. In this there was also a rebirth of the Athenian ideal of philosophy as a study unto itself, and with the huge expansion of the universities and the patronage of the royal courts there was a way to pursue philosophy independent of theology and Church support. Natural philosophers were still called upon to justify their enterprise, something they did by calling attention to its civic and state utility.

Although most humanists were more concerned with understanding the books of the Bible and Cicero in their original languages than in predicting the paths of planets, their enterprise helped infuse new life into natural philosophy. Humanism did this in three ways: humanists rediscovered and translated classical scientific sources from the original Greek; humanist methodology treated written sources in a more skeptical manner; and humanism introduced a new purpose for, and mode of, scientific discourse.

Equally important, humanism revived Aristotelianism, both by rediscovering early Greek versions of Aristotelian texts formerly known only through Arabic translations and by forcing scholastics to make their arguments and methodology more rigorous. As a result, Aristotle's system did not give way before the humanist onslaught; rather, it incorporated much of the methodology and rigour of humanistic studies while retaining its basic framework. The Aristotelian system had proven extremely fruitful as a research program, since it provided an all-encompassing study of the physical world including physics, astronomy, and biology, and of the spiritual and social world using metaphysics, logic, and politics. Until an equally sophisticated paradigm could be established in the seventeenth century, Aristotelianism remained useful and necessary. Thus, the history of natural philosophy throughout the fifteenth and sixteenth centuries is one of the refinement and triumph of Aristotelianism, rather than of its defeat.

Two major factors contributed to the rediscovery of Greek natural philosophy in this period. The first was the fall of Constantinople to the Turks in 1453. Before this date, individual Greek manuscripts were traded from Byzantium or were discovered in various Italian monasteries. But with the fall of the last outpost of ancient Greek scholarship, hundreds of books, some of them literally thrown over the walls to save them from the invaders as the city fell to the Turkish army, were brought all at once to Italy. Knowledge of Greek now became absolutely necessary for scholarly work. The flooding of the intellectual market with Greek texts coincided with the second impetus to the rediscovery of Greek natural philosophy. This was the patronage of the Medicis, who were interested in a full translation of Plato. Cosimo de' Medici, head of a powerful Florentine banking family, became interested in the metaphysical philosophy of Plato in 1439 and by the 1450s encouraged humanists such as Marsilio Ficino (1433–99) to undertake translations of his work. Cosimo set up the Platonic Academy, with Ficino as its head, and in a relatively short time this group translated many of the important works of Plato into Latin. This rediscovery, combined with the discovery of mystical and magical treatises such as those of the supposed Hermes Trismagistus and the Jewish cabala helped to develop Renaissance magic as a much more esoteric study than its practical medieval counterpart, as seen in the *Book of Secrets*.

What made the rediscovery of Greek natural philosophy, and with it the growth in interest in the study of nature, a European phenomenon, rather than just an Italian one, was the invention of the printing press. In 1448 Johannes Gutenberg (c. 1397–1468) introduced movable-type printing, thereby revolutionizing communication. Movable type printing was not in itself a revolutionary idea, but it represented the perfection and combination of a number of existing technologies. Printing, using carved wooden blocks, had been around for over 1,000 years and was used by the Chinese from around 1045. The Chinese inventor Bi Sheng (990–1051) created a movable-type system using porcelain characters, but it is not clear if knowledge of Chinese printing was known in Europe. Block printing was in use in Europe by the beginning of the fifteenth century. Despite the invention of most of the components for printing in China, the development of printing for publication in that country was inhibited both by the pictographic nature of the language, which would have required thousands of characters, and by the threat it posed to the monopoly on writing of the established class of scribes. By contrast, Gutenberg worked with only 24 letters (the use of "j" and "u" had not been standardized), plus capitals, punctuation, and a few special symbols, at a time when scribes in Europe were scarce and expensive.

Gutenberg combined two Asian inventions, the screw press and paper, to develop his movable-type printing press. Paper had been invented in China around 150 BCE and was manufactured in Europe by 1189, offering a less costly alternative to vellum and parchment. Gutenberg created typographic characters by scribing each individual letter into a hard metal (steel), then using these as a punch to make a set of moulds out of a softer metal (copper). He could then cast as many letters as he needed out of a lead alloy. The letters were uniform in size and shape and could be assembled and printed, then separated and recombined repeatedly.

Gutenberg's work was meticulous, since he was attempting to replicate the typography of the written manuscript. This attention to detail and the cost of creating the actual press led him to seek financial backing from Johann Fust of

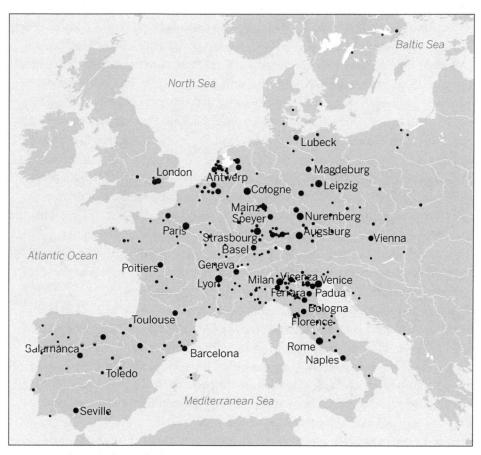

4.1 SPREAD OF PRINTING IN EUROPE

The centres of printing between 1452 and 1500.

Mainz in 1450. Gutenberg's project was the "42-line Bible" (also known as the Gutenberg Bible or the Mazarin Bible). He was not as good a businessman as he was an engineer, however, and lost much of his equipment to Fust to pay his debts, who completed the printing in 1455. About 300 copies of the Bible were printed and offered for sale at 30 florins each, which was equal to about three years' wages for a clerk.

Gutenberg's press was copied by many others, and by 1500 there were more than 1,000 printers working in Europe. (See figure 4.1.) Bibles, religious texts, and indulgences were in high demand. Indulgences were slips of paper that could be purchased from the Church for the remission of sins, and they had been used by the Church to finance everything from the Crusades to building cathedrals. With the aid of the printing press huge numbers of indulgences could be produced, sometimes as many as 200,000 in a single print run. The demand for other kinds of books also exploded as everything from Greek and Roman literature to medical texts became available. The humanist interest in first Latin and then Greek literature supplied materials for the printers and also created a demand for this classical material.

The effect of printing was enormous. Books were now available to people who had never seen a manuscript of any kind. Printing made information far more widely available, and a huge storehouse of material was opened to a growing audience. As the cost of books declined, more people could afford to own them, and reading habits changed as literacy spread. With the introduction of page numbers (which familiarized readers with Arabic numerals), tables of contents and indexes became possible. This meant that a book did not have to be read from cover to cover but could be dipped into just for information the reader thought was pertinent. Since people could now potentially own many volumes, they could compare one text with others, an impossibility in a scribal age.

The media scholar Marshall McLuhan and others have argued that the introduction of mass printing technology ultimately changed the very psychology of Western society. The change from a non-literate to a literate society changed the sense of time and space, shifted the locale of truth from human memory to written records, degraded memory, promoted dissent and a wider world view, and was partly responsible for the development of the concept of professionalization and the creation of the "expert."

In the realm of natural philosophy the introduction of print changed the discourse as well. Printing helped establish the definitive and corrected version of Greek and other natural philosophy texts, since several manuscripts could be compared for the most authoritative version. This prevented scribal drift, or the compounding of simple errors such as spelling mistakes that grew worse with repeated copying. It also allowed the insertion of illustrations, charts, and maps, items that had been so prone to scribal errors that they had usually been omitted from manuscripts or were useless. This meant that scholars could concentrate on finding new knowledge rather than constantly correcting the old. Because readers could purchase and borrow numerous books at relatively cheap prices and without personally going to monasteries housing the original manuscripts, a search of the literature was possible, as were comparisons of alternative versions, especially of star charts or botanical illustrations and descriptions. New information could also be disseminated rapidly. For example, news of Christopher Columbus's voyage in 1492 was printed immediately on his return to Spain and translated from Spanish into German, Italian, and Latin within the year, while knowledge of Marco Polo's thirteenth-century visit to China was known only to a select few, even into the fifteenth century. Finally, printing provided natural philosophers with paper calculational devices, a public forum for their ideas, and a republic of letters within which to converse with people of similar interests and aptitudes.

Copernicus, Tycho Brahe, and the Planetary System

Probably the most famous natural philosopher influenced by humanist ideas, for whom the printing press transformed his research and dissemination, was Nicholas Copernicus (1473–1543). Copernicus was born in Torun, Royal Prussia (then and now part of Poland),[1] a relatively isolated intellectual outpost. He travelled as a student to Italy, where he learned humanistic techniques and consulted original manuscripts. The most significant documents he found there were complete copies of Ptolemy's *Almagest*, the single most important source for astronomy at the time. No complete manuscript of the *Almagest* was available in all of Royal Prussia in the 1480s. By the time of his death in 1543 there were three different editions of Ptolemy's book in print, allowing Copernicus and other astronomers to compare astronomical tables, discover the discrepancies between ancient observations, and establish a new model. Copernicus also had a printed version with diagrams of Euclid's *Elements* (Venice, 1482) and a list printed by Johannes Regiomontanus (1436–76), who had printed the first edition of Ptolemy's *Almagest*, of all the important scientific works from antiquity, which became the required reading list for sixteenth-century astronomers and mathematicians. In Italy, Copernicus also encountered manuscripts that originated in Arabic sources. Historians have now shown that Copernicus's ideas owed much to these Islamic astronomers, and that his new planetary model was a part of a conversation across cultures, rather than a result of one solitary thinker. Copernicus did not care where his concepts came from but was happy to try a number of strategies to see what would work.

When Copernicus studied Ptolemy's astronomy and compared it to medieval star and planet charts, he saw serious problems. Not only did the predicted locations of the celestial bodies differ, but Copernicus believed that Ptolemy had violated his

---

1. Royal Prussia became subject to the authority of the Polish Crown in 1454 but became part of Prussia in 1772. It returned to Poland after World War II.

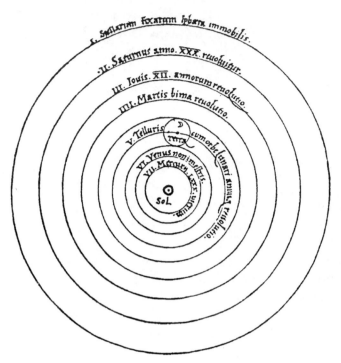

4.2 THE COPERNICAN SOLAR SYSTEM
FROM *DE REVOLUTIONIBUS* (1543)

own insistence on perfect circular motion in the heavens. Copernicus decided, as a mathematical exercise, to reverse the heavenly arrangement and place the sun at the centre with all the planets, including the Earth, revolving around it. In this schema, the sun remained stationary in the centre and the Earth now had a diurnal (daily) motion in order to account for night and day, as well as an annual orbit around the sun. (See figure 4.2.) To this, Copernicus added a third motion of the axis of the Earth's rotation that accounted for the seasons and the annual inclination of the zodiac.

Copernicus's system was just as complicated mathematically as the Ptolemaic system had been, but it did explain a number of anomalies that had been worrying astronomers for some time. For example, there was no good explanation in the Ptolemaic system for why Mercury and Venus never appear more than 45° away from the sun. Copernicus's system solved this issue by placing the inner planets between the Earth and the sun. In addition, the heliocentric model resolved the major issue of retrograde motion, which had led Ptolemy to devise the epicycle.

Moreover, Copernicus's system was aesthetically pleasing and eliminated the diurnal motion of the whole universe. It was not without its own problems, however. For example, Venus and Mercury should have phases like the moon in this new schema, but these had never been observed. More worrying, the stars did not appear to move, even though Copernicus's schema called for the Earth to move across the skies. The astronomers of the day assumed that the stars were close enough to the Earth that the angle they were viewed at would change if the Earth was orbiting the sun. This is called parallax and was not seen until 1838. Given the vast distance to the stars, it was not measurable until the development of powerful telescopes.

If the Earth was actually moving with a triple motion as Copernicus suggested, other questions of a more terrestrial nature might be asked. Why could birds fly east? Why did balls fall straight down? Why couldn't we feel the Earth moving? There existed no test that could demonstrate the motion of the Earth, and this flaw plagued astronomy for several generations.

Above all, Copernicus's system violated the whole Aristotelian ordering of the universe. Without the Earth in the centre, Aristotle's physics of "natural motion" fell apart. Catholic theology had come to depend both on Aristotelian explanation and, especially, on the centrality of the Earth as the least perfect part of the universe and therefore at the same time both the site of sin and transgression and the focal point for salvation. If the Earth was just one of many planets, could there not be other Christs and other salvations? For just such speculations, Giordano Bruno (1548–1600) was burned at the stake in 1600. Therefore, it is not surprising that Copernicus, a canon and thus an officer of the Church, delayed publishing his

Large armillary (c. 1585).

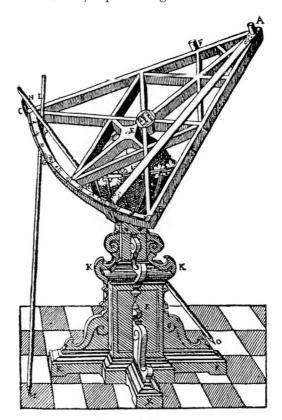

Sextant (c. 1582).

4.3 TYCHO BRAHE'S OBSERVATIONAL EQUIPMENT

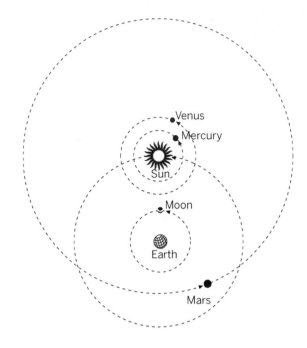

4.4 TYCHONIC SYSTEM

ideas until he was on his deathbed. He agreed, reluctantly, to the publication of *De Revolutionibus Orbium Coelestium* (*On the Revolutions of the Heavenly Spheres*) in 1543, through the persuasion of his friends, especially Georg Joachim Rheticus (1514–74). Rheticus was overseeing the publication of Copernicus's work until he was forced to leave Nuremberg. Andreas Osiander (1498–1552) took over and added an unauthorized preface claiming the whole thing was only a hypothesis. Hypothesis or no, the printing of *De Revolutionibus* allowed the whole European scientific community to learn of Copernicus's ideas, and a century of controversy began.

Scholars who read Copernicus's work fell into two categories: philosophers who were interested in the overarching cosmology and mathematicians who wanted to use the calculations without worrying about the model. Many philosopher/astronomers wished to tinker with the model in order to make one more acceptable to the Church. Tycho Brahe (1546–1601) was probably the most prominent of these. Tycho was a Danish nobleman who, rather than serving as a military commander to his king, as was typical of someone in his social position, offered his heroic astronomical work as his feudal dues instead. He was deeply indebted both to the humanist rediscovery of ancient natural philosophy and to the technology of the printing press. Even more than Copernicus, Tycho was able to compare printed tables. He was, indeed, a self-taught astronomer, learning his craft initially from printed books. He was also the best naked-eye observer in Europe. He built a huge observatory and the largest pre-telescopic astronomical instruments ever seen. (See figure 4.3.)

Tycho devised a planetary system, often called the Tychonic system, that was halfway between those devised by Ptolemy and Copernicus. In this system, the sun and moon revolved around the Earth, while everything else revolved around the sun. This saved the Earth as the centre of the universe and of God's grace while also explaining the problems of Mercury and Venus. (See figure 4.4.)

Using his impressive astronomical equipment Tycho also made some of the most important comet and new star sightings of the sixteenth century. He observed the comet of 1577, for example, and showed that its path sliced through the orbits of other planets. This was a major discovery, since it forced people to think about the physical reality of the solid transparent spheres of Aristotelian cosmology. Where did these comets come from? How could they be imperfect (transitory) and yet supralunar (above the orbit of the moon)? Tycho and others proved that their paths were supralunar and thus discredited the traditional physical explanations of the universe. But Tycho had no alternative physics to propose, which may help to explain the reluctance of astronomers and natural philosophers to abandon Aristotle.

Another of Tycho's discoveries was the sighting of several new stars—stars appearing and continuing in the skies where none had been before. Again, this made a case against the unchangeability of the heavens, and, because Tycho's observations were so good, the new stars could not be ignored. The sighting of the new star of 1572 was, in fact, a completely different event than earlier supernovas, since many people, following Tycho's lead, were able to observe the phenomenon simultaneously and report within the year to the academic community. This was one of the first instances of community agreement rather than scholarly authority as the basis for establishing a scientific "fact." The making of scientific facts increasingly became a public enterprise.

Some historians have pointed to Copernicus's work as the beginning of the scientific revolution or at least a Copernican revolution in astronomy. If by revolution we mean a rapid shift from an old to a new model, it largely did not happen. Despite Copernicus's radical reordering of the universe and Tycho's impressive observations, people were reluctant to abandon the Ptolemaic system and embrace Copernicanism. It was never fully accepted until the heliocentric schema was modified by Kepler and Newton, and only in the late seventeenth century did it become the generally accepted model. During the sixteenth century mathematical astronomers took up the technical aspects, philosophers the descriptive. Still, a number of astronomers saw the benefit of at least considering this new framework, and gradually—at different times in different places—the sun was accorded its place in the centre of the universe. Astronomers' reasons for moving from one system to the other were complex. The historian Thomas Kuhn argued, for example, that some people found the Copernican system aesthetically pleasing. A few isolated thinkers such as Englishman Thomas Digges (1546–95) and German Michael Mästlin (1550–1631) accepted Copernican cosmology, while others, such as the close-knit scholarly community at Wittenberg, adopted a hybrid system very early in the 1550s. Galileo's championing of Copernicanism had much to do with patronage, as we shall see, although Galileo was also influenced by aesthetics and concerns with bringing astronomy and physics into accord.

The Age of Exploration

The debate about the correct model of the heavens was not just a scholarly squabble. All over Europe rulers and entrepreneurs had an urgent need to understand and predict the motions of the skies, since increasingly they were interested in long sea voyages of trade and discovery. From the Crusades on, Europeans had been interested in the exotic goods available through trade with the Middle East and Asia. By the fifteenth century this trade was completely controlled by the Turkish Empire, especially after the fall of Constantinople, and so enterprising European nations decided to circumvent the bottleneck of the Bosphorus and go around the middlemen. The Portuguese began by coasting down Africa and found, despite the closed Indian Sea depicted in Ptolemy's famous maps, that they could reach the East via the Cape of Good Hope, although they still had to pass through waters controlled by Islamic people.

Islamic traders had been sailing extensively in the Indian Ocean for many years before the Portuguese started their project of exploration and trade. Arabic traders moved between the coasts of Africa and India from at least the twelfth century, trading, establishing outposts, and interacting with the Indian population. The Portuguese were interested in voyaging for a complex mixture of reasons, including curiosity and imperial expansion, but most especially commercial concerns. They were very interested in working with Arabic traders in order to develop new routes to the East. The Portuguese were happy to use any information they found and often appropriated the techniques, maps, and matters of fact from other travellers in the region. Their maps of the Indian Ocean, for example, made use of the knowledge of Arabic traders, veterans of those waters for several hundred years. When Europeans published these maps, they erased the Muslim sources, leaving behind a tale of heroic European adventure.

The Chinese had also developed the necessary navigational and mapping skills to undertake significant oceanic voyages, long before the Europeans started their "age of exploration." Most famous are the voyages of the imperial eunuch Zheng He

(1371–1433/35). Zheng He was born into a Muslim family of the Hui people in Yunnan. When Yunnan fell to the Emperor's forces, Zheng He was captured and castrated. He became a powerful member of the imperial court under the Yongle Emperor, who sponsored seven naval expeditions, with Zheng He as the admiral. These voyages, with hundreds of ships and tens of thousands of troops on board, sailed all through the Indian Ocean to the Horn of Africa and Arabia. (The theory that Zheng He actually sailed around the world is without foundation, however.) Zheng He sailed from 1405 to 1433 (although he may have died on the final voyage), bringing gifts to the leaders he contacted and returning with tributes to the emperor. Most famous was the giraffe he brought back from Africa during his third voyage of 1413–15. Zheng He's achievement was considerable, but it is important to note that he followed long-established and well-mapped routes, some dating to the Han dynasty. For example, when his fleet arrived in Malacca in 1407, there was a sizable Chinese community already established there.

Historians have debated why the Chinese did not continue with this program of exploration after Zheng He. It is clear that the death of the Yongle Emperor was a key factor, since his successor immediately stopped the voyages, which he saw as expensive and unnecessary. It also seems likely that this was a political issue, since Zheng He's achievements represented the power of the eunuchs over the scholar/bureaucrats who were less interested in these voyages. The Chinese began to concentrate more on domestic issues and were less interested in contact with the wider world, although they continued to trade along the Silk Road and interact with other peoples in the China Sea.

Others who might have had the technical ability to sail long distances were the peoples of the Americas. Jacob Bronowski argued that the "new world" did not travel out to the old world because it lacked a sense of the heavens as a wheel—an invention little used by the Maya or other South American civilizations. While this may have contributed to a lack of exploration, two simpler reasons restricted Mayan scientific activity. The first was a series of collapses caused by endemic warfare and agricultural failure because of drought and environmental degradation. Mayan social structure did not adapt well to the challenges, making the situation worse. They did not have the time or periods of peace to develop natural philosophy, and they never separated the study of nature from their religious practices. The second problem was technological. The Maya had great mathematicians and engineers, but they did not master a number of technologies, especially high temperature smelting or glass-making, leaving them with Neolithic tools.

While this new "age of exploration" had little influence on the Chinese or Mayan world view, this was a critical period in the development of European consciousness. Although the Chinese had sailed farther and many fisher folk had been traversing the Atlantic Ocean for centuries, the achievements of Vasco da Gama (c. 1469–1524) and Christopher Columbus (1451–1506), as well as those who followed, fundamentally changed the way Europeans understood the Earth and their relationship to it. These early explorers, equipped with a Christian and imperial belief in the righteousness of their cause and the superiority of their understanding, challenged the authority of the ancients, especially Ptolemy. Ptolemy's *Geographia* had only been rediscovered by humanists in 1406, providing another view of the globe that could be used and challenged. As with other natural philosophical endeavours, then, humanist rediscovery sparked an extension and eventually refutation of ancient knowledge of the globe. Columbus and those who came after demonstrated to Europeans the existence of a continent completely unknown to the ancients (though familiar to its inhabitants). More importantly for natural philosophy these explorers disproved a number of ancient and medieval theories of the Earth, most particularly by demonstrating that the globe had a much larger proportion of dry land than had hitherto been suspected, that it was possible to sail through the equatorial regions without burning up, and that people could and did live south of that equatorial region in the lands known as the antipodes. Columbus did not prove the world was round—this had been known by learned men since antiquity—but he did prove that the globe was navigable and, ultimately, exploitable by Europeans.

The prime motivating factor for these voyages was amassing great wealth, both for the individual and for the country sponsoring the enterprises. At first, the destination was the Far East—Cathay and the Spice Islands. The Portuguese were most successful at reaching these areas, setting up key trading depots in Goa (India), Malacca (Malaysia), and the Moluccas (Spice Islands). The Spanish, having reached the Americas by mistake, soon modified their mission, and although they continued to seek gold and especially silver, the *conquistadors* began to focus on colonization, seeing the natives as a useful slave population and one that could be converted easily to Christianity. Later, as the cultivation of sugar and cotton became more important to the Spanish, the African slave trade was introduced into this economic arrange-ment. And yet it would be a mistake to separate these imperial and mercantile enterprises from the growing interest in and study of the Earth. The study of nature was inexorably linked with religious and mercantile concerns. The "discoveries" of this age of exploration encouraged new innovations in cartography and navigation, led to a changing understanding of the terraqueous globe, spawned an interest in the effect of climates on human beings, and launched ethnographic investigations and debates concerning the New World peoples.

There was a burgeoning interest in the mapping of the world in the sixteenth century, undoubtedly influenced by that fifteenth-century rediscovery of Ptolemy. At first, charts and plots were used as aids to descriptive and experiential knowledge, but eventually European rulers, investors, and scholars wanted to visualize their world in this new graphic way. Countries such as Spain and Portugal were quick to develop state-controlled repositories of navigational maps and charts. Later, monarchs called for the mapping of their individual countries and regions, as well as creating larger maps of imperial concerns. The result was a flood of map production, including world atlases by Gerhard Mercator (1512–94) and Abraham Ortelius (1527–98), beautifully engraved in the Netherlands, and country surveys by Christopher Saxton (c. 1542–1611) in England and Nicolas de Nicolay (1517–83) in France. Working in Amsterdam Willem Jansoon (1571–1638) and his son Johannes Blaeu (1596–1673) produced a series of detailed world maps. (See plate 2 for Blaeu's 1664 map.) Maps became objects of desire for prosperous merchants, as we can see from numerous Vermeer paintings of merchant houses with beautifully coloured maps hanging on the walls. They were used to visualize and control space, to build empires, and to swell local and regional pride and identification.

One of the most troublesome aspects of the New World discoveries was the fact that there were people there. Who were they? What were they? While European scholars and explorers could use only European categories and under-standing to interpret what they encountered, this contact with a previously unknown Other had far-reaching implications for European thought. Early explorers interpreted the customs and behaviours of those they encountered from a European viewpoint and tried to eliminate customs that did not suit their preconceptions, such as the lack of private property or a nomadic way of life. Sixteenth-century Spanish theorists tried to fit Amerinds into the only classifica-tion system they knew: Aristotle's. Thus, men such as Bernardo de Mesa argued that the Amerinds were natural slaves. The discovery of the Incas and Aztecs in the 1520s made this harder to believe. Clearly, in Aristotelian terms, these people were civilized. They had government and infrastructure and lived in a complex community. And so thinkers such as Francisco de Vitoria (c. 1492–1546) claimed that these people were natural children, based on the idea that they made category errors, such as engaging in cannibalism, bestiality, or eating dirt, but had the capacity to learn from their mistakes. They had to be protected because with training they might be raised up to adult (that is, European) status. This opinion was never shared by the majority, since it implied that eventually these children would grow up and would have to have their property restored to them. Another minority opinion, that of Michel de Montaigne (1533–92), had far-reaching effects. Montaigne argued that the Timpinambas of Brazil, although cannibals, were a noble race, more moral than Frenchmen, even if they did not wear trousers. This idea of the noble savage recurred most famously in the writings of Jean-Jacques Rousseau.

Paracelsus, Medicine, and Alchemy

Internal trade in Europe had been growing steadily from the time of the Crusades and by the sixteenth century had developed into a strong mercantile culture and economy. Greatly expanded by the gold and silver bullion flooding into Europe as the New World trading networks developed, manufacturing and trade among European nations expanded considerably. Mining in Europe and the New World became a growth industry, and with these economic and industrial changes, concomitant developments occurred in natural philosophy, especially in theories of mining and metallurgy on the one hand and alchemy on the other. As well, the increasing numbers of skilled artisans began to develop links with natural philosophers, asking new questions and developing new systems of investigation.

Mining of precious metals and other minerals had taken place since antiquity, but the demand for these goods soared in the sixteenth century. Coal for heat, iron for steel, tin and copper for manufacturing were all profitable minerals. There were a number of technological problems to be overcome in mining these substances, not least the water present in mines of any depth. Pumps were devised, although none were completely satisfactory. The refining of metals was also a process that had to be worked out, and Georgius Agricola (1494–1555) in *De Re Metallica* (*On the Nature of Metals*, 1556) was the first to explain some of these processes in natural philosophical terms. (See figure 4.5.) Agricola was humanist-trained and, clearly from his use of Latin, was interested in introducing the study of metals to a scholarly audience. On the other hand, he lived in Bohemia and Saxony, the richest mining lands in Europe, and he married a mine owner's daughter, so he was not exactly a disinterested party.

Mining produced serious illnesses among the miners, so it is no surprise to find a physician who interested himself in these cases. The German physician Theophrastus Bombastus von Hohenheim, known as Paracelsus (1493–1541), was influential in bringing together medical and alchemical knowledge, and he is recognized as one of the main creators of iatrochemistry or medical chemistry. His life was deeply influenced by the religious and social crises in the German states. Paracelsus was born in Zurich; his father was a physician who wanted him to follow in the profession. In 1514 he spent a year working at the Tyrolian mines and metallurgical shops of Sigismund Fugger, who was also an alchemist. It was through Fugger that Paracelsus became intrigued by the nature of metals, and he spent much time during his life trying to identify and discern the properties of metals. After he left Tyrol, he travelled widely across Europe, studying briefly with alchemists in France, England, Belgium, and the countries of Scandinavia

4.5 ORE PROCESSING EQUIPMENT FROM AGRICOLA'S *DE RE METALLICA*

before finally going to Italy, where he claimed to have earned a medical degree in 1516 at the University of Ferrara.

In 1526 Paracelsus settled in Strasbourg to practise medicine. He treated miners' diseases, especially black lung. His alchemical work led him to become an advocate of the use of metals rather than traditional plant-based drugs in treatment. Most famously, he prescribed mercury for cases of the new disease of syphilis, a cure only slightly less excruciating than the original symptoms! Paracelsus's fame grew, and when the printer and publisher Johann Froben of Basel fell ill and local physicians failed to cure him, he sent for the young doctor.

Paracelsus cured him. At the time, Desiderius Erasmus, the famous Dutch humanist and biblical scholar, was staying with Froben, so Paracelsus's success was widely noted. Paracelsus was offered the position of City Physician and Professor of Medicine in Basel. He accepted, but held the position for only two years, since his radical ideas about the treatment of disease caused great controversy. He started his career as City Physician by publicly burning copies of Galen and Avicenna in order to demonstrate his rejection of the old medicine, which treated diseases with herbs. He was also radical in other ways, insisting on lecturing in German rather than Latin. He was loved by his students but hated by his associates, whom he frequently criticized.

The city officials defended their choice of Paracelsus against a clamour of protest from apothecaries and other doctors. Then the Canon Lichtenfels fell ill and offered 100 gulden to any doctor who could cure him. Paracelsus used his metallic system and Lichtenfels recovered but then refused to pay. Paracelsus took him to court, but either because the fix was in or because of some legal mistake on Paracelsus's part, he did not win his case. He left his position and spent the remainder of his life wandering through Europe, repeatedly running into trouble with authorities for his radical ideas. Lacking a powerful patron to protect him, he was in constant danger of being arrested by secular authorities or accused of heresy or witchcraft by religious officials. Finally, in April 1541 he found employment at the court of the Archbishop Duke Ernst of Bavaria. Ernst was very interested in alchemy, so it was likely the patronage position was offered for both medical and alchemical reasons. Unfortunately, Paracelsus, weakened by years of hardship, died in September that same year.

Unlike Aristotle, who had argued that there were four basic elements, Paracelsus and many of his fellow Renaissance alchemists claimed there were only three: salt, mercury, and sulphur. The careful combination of these three, with arduous, secret, and prolonged laboratory manipulations, might lead to the illusive Philosopher's Stone, the source of eternal life, the gold of the soul, and perhaps material gold as well. While Paracelsus can be seen as an alchemist, he was not really interested in transmutation. Instead, he was interested in iatrochemistry—medical chemistry. He shared the slowly evolving view that alchemy should be concerned with employing the material world for useful purposes, not with the fruitless effort to create precious metals. Although much of his work had mystical aspects, he also promoted the concept of understanding matter based on elemental composition, one of the foundational ideas of later work in chemistry.

Patronage and the Study of Nature

There was often no clear line to be drawn between the esoteric research of the alchemist and the mundane concerns of the apothecary. Both belonged to a

4.6 MAJOR SITES OF NATURAL PHILOSOPHICAL WORK IN EUROPE, 1500–1650

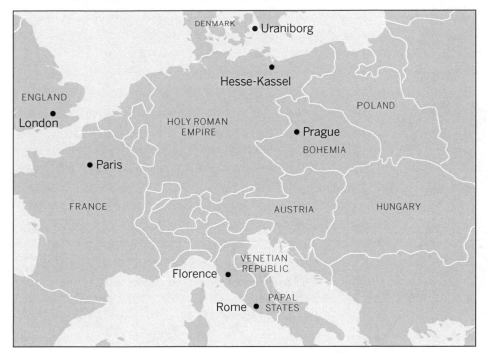

growing group of skilled artisans who plied their trade in increasingly large numbers in the urban centres of sixteenth-century Europe. Printers, instrument makers, surveyors, and shipwrights all began to ask questions about how the natural world could be used to their benefit. They often used and sometimes taught mathematics. This community of superior artisans, together with scholars trained and employed in non-traditional settings such as courts or the homes of private patrons, developed new questions about the make-up, design, and running of the world that would lead, by the next century, to a major reorientation of the scientific enterprise.

One place where natural philosophers, mathematicians, and practitioners came together was the princely court. During the Renaissance these were sites of spectacle and culture where political, cultural, and intellectual patronage encouraged some of the most glittering and opulent courts seen since antiquity. The earliest of these courts were, of course, Italian, and as we have already seen, the Medicis of Florence gathered together some of the foremost artists, humanists, and natural philosophers of their day. Other princes and courts followed suit, and

CONNECTIONS

Patronage and
the Investigation
of Nature: John
Dee and the Court
of Elizabeth I

The life of the famous necromancer, mathematician, and natural philosopher John Dee provides a fascinating glimpse into the complex and sometimes dangerous world of patronage. Dee worked hard but ultimately unsuccessfully to gain a place in Queen Elizabeth's court as her Royal Philosopher; in the process he pursued a number of practical projects that took him far from the philosophical work he valued. When he insisted on the importance of his scholarly work, his patron became less and less interested and he received less support.

John Dee received his education at Cambridge and very soon established his superior understanding of mathematics and geography. He went to Louvain to study with Gemma Frisius and Gerard Mercator, two prominent mathematicians and globe makers, and when he returned to England, he set himself up in London as an astrologer and geographical advisor. Many explorers, such as Humphrey Gilbert, asked his advice about navigational and geographical issues, including the question of the existence of a northwest or northeast passage. Dee became astrologer to the princesses Mary and Elizabeth. He was charged with treason for casting Mary's horoscope and appeared before the Star Chamber (a special law court often dealing with political trials), but was eventually able to clear his name. After Mary's death, he became Queen

soon natural philosophy became part of this patronage system, affecting the topics of investigation and how they were investigated. Hans Holbein's portrait entitled *The Ambassadors*, painted in 1533, demonstrates the importance of mathematical instruments to the self-fashioning of courtiers. (See plate 3.) These two men, French ambassadors to the court of Henry VIII, display a celestial and terrestrial globe, a quadrant, a torquetum, and a polyhedral sundial as evidence of their learning and wealth.

Patronage was a system of dependency, with personal contracts between two individuals: the patron and the client. The patron had power, money, and status, but wanted more. The client could give the patron more of these while getting some for himself. It was thus a two-way and often volatile relationship. The whole system was based on changing the balance of status. In natural philosophical relationships the client claimed special knowledge or skill, usually with some practical application, although sometimes he simply offered the patron the prestige of being able to surpass the knowledge of some other prince's natural philosopher.

Elizabeth's astrologer, advising her on the luckiest day for her to hold her coronation. Elizabeth took his advice and consulted him on many matters of astrological, geographical, and imperial importance.

Dee, however, had his sights set on higher goals. He sought to understand the underlying basis of matter through alchemy and the universal language of creation through Hermetic philosophy and magic. He hoped to develop a completely new philosophical structure for understanding the world, one that would lead to a unity of all mankind, but he could not persuade the queen to provide him with the necessary stipend that would have allowed him the freedom for such work. He sought the sort of fame and stability that Johannes Kepler had as Imperial Mathematician to Rudolph II. But Elizabeth was both practical and cheap, and although she gave Dee gifts, as was appropriate in a patron–client relationship, she never conferred on him the money or title that he sought.

Dee then moved to more esoteric and supernatural research, searching for the transcendent understanding of divine forms (the Platonic Ideals of nature) through scrying (crystal ball gazing) and angelic and demonic communications. His lack of success at Elizabeth's court caused him to try his luck at various other European courts. Unfortunately, the Polish courtiers he met were suspicious that he was an English spy, and he fared little better in Bohemia; he was forced to flee Rudolph's court, and on his return home to England, discovered his house and library had been vandalized. Although Elizabeth gave him a small position as Warden at Christ's College, Manchester, Dee was never able to achieve the success he had earlier in her reign. He died in poverty, with only his daughter to care for him. Dee's story, then, is a cautionary tale about the dangers as well as the rewards of scholarly patronage.

The philosopher sought to gain the attention of the would-be patron by dedicating a book or sending a manuscript to him or her, by circulating a letter concerning the patron's interests, or by publishing a book acknowledging the patron's greatness. Through negotiations the patron granted some court or household position to the scientist. This generally led to science at the courts that was useful, daring, and often controversial. In some cases cooperative enterprises were undertaken with the patron or other members of court.

There are numerous examples of these client–patron relationships, including that of Leonardo da Vinci (1452–1519) at the court of Charles VIII of France; German prince-practitioners such as Rudolph II and Wilhelm IV, Landgraf of Hesse; and the astronomer and mathematician John Dee (1527–1608) at the English court of Queen Elizabeth I. The best example of the patronage relationship and its effect on natural philosophy, however, is the life of Galileo Galilei (1564–1642). While modern commentators remember Galileo's final condemnation by the Roman Inquisition, he was famous in his day for his telescopic sightings. Through his astronomy and even more through his physics, Galileo constructed an abstract mathematical schema, suggesting the abstraction and mathematization of the world so integral to early modern natural philosophy. He believed that God had constructed the world using number, weight, and measure, and thus he replaced the study of causes with the study of laws. He used measurement and experiment, usually seen as part of modern scientific method. But perhaps what is most interesting about Galileo is that he did all this not within theological institutions as Copernicus had done, or in the universities as Newton would do, but at court. Galileo was every inch an early modern courtier, a kind of intellectual knight, with power to gain (and lose), and constantly looking for innovations to aid and glorify his patron.

Galileo

Galileo was born in Pisa in 1564. He moved to Florence early in his life and always thought of himself as a Florentine. His father, Vincenzio Galilei, was a famous musician who discovered a number of important mathematical musical laws. Vincenzio wanted his son to become a physician who, he said "made ten times as much money as a musician," but Galileo was more interested in mathematics. His first job, at the University of Pisa, was as a teacher of mathematics, at the bottom of the academic status ladder. His first significant post was at the University of Padua, which was under the control of Venice, and he used his patronage connections with powerful people in the Venetian elite to work his way up. Galileo was always in need of money, because he had seven brothers and sisters who relied on him for support. He needed to find big dowries for his sisters, and his errant brother Michelangelo Galilei was constantly in debt.

While at Pisa, and especially at Padua, Galileo began to study motion, although he did not publish his findings for 40 years because he could not figure out all the details to his own satisfaction. After many years he decided this was not an important question; rather than look for the cause, he developed laws of *how* motion worked. He eventually published his mechanics in *Discourse on the Two New Sciences* (1638). He rejected Aristotelian notions of motion, showing that speed does increase continuously, at least in free fall and, therefore, that impetus (the force impressed on an object, which Aristotle said would wear out with time) did not exist. Instead, Galileo argued that continuous motion once imparted, or continuing stillness, would remain forever. What separates this from Newton's later idea of inertia is that for Galileo continuous motion was circular. In Galileo's system a ball set in motion on the Earth, if unimpeded by friction or any other extraneous force, should travel continuously in an orbit around the Earth. Probably Galileo's most significant achievement in mechanics was his development of a clear picture of abstract and measurable motion.

For many years historians believed that Galileo did only thought experiments. We now know that he did practical experiments, although probably his most famous one, the Leaning Tower of Pisa experiment, was not performed by him, making this the most famous unperformed experiment in the history of science. It is possible that one of his students dropped two balls of different masses from the top of the tower, although it is not clear that Galileo witnessed this. The point of the experiment was to find out whether Aristotle, who predicted that the two balls would fall at different rates proportional to their weight, was correct. Galileo's actual experiments on motion led him to predict that the two balls would fall at the same rate. According to Galileo, if the two balls were dropped from the tower at the same time, they would hit the ground at the same instant. Allowing for a small variation due to air resistance, Galileo was correct.

The problem with the study of falling bodies was that they travelled far too fast for quantitative analysis with the equipment available at the time. There were no stopwatches in his day, so Galileo devised a method to "dilute" the rate of free fall. He rolled balls down an inclined plane, which had small notches at regular intervals. He measured the time by listening to the click as the ball hit these bumps and comparing it with someone singing Gregorian chants. He discovered that distances from rest were proportional to the squares of the elapsed time ($k = d/t^2$). This was a huge discovery, achieved by removing all real-life distractions, thereby creating an almost frictionless plane on which he could study an ideal example. Galileo was no longer asking *why* bodies fell (the cause) but rather measuring how fast they did so. This law was extremely influential for Newton's later work as he applied it to the universe and constructed a world subject to even more accurate measurement. Newton, like Galileo, avoided the question of causes.

Galileo also worked on the problem of projectile motion, important to the princes and principalities for whom he worked, since it was connected to ballistics and warfare. He determined that cannonballs move in a parabola by dividing their motion into two parts (forward motion and earth-seeking motion). He discovered that a ball shot from a cannon will hit the ground at the same moment as one dropped from the same place and determined that pointing the cannon at a 45° angle produced a maximum range. (See figure 4.7.) Since Galileo was a Copernican, he used this argument of the different vectors of motion to argue for the movement of the Earth.

As Galileo worked his way up the patronage system, he found that astronomical work was more successful in attracting patronage than mechanics or mathematics. While ballistics had been useful, it was the telescope and the discovery of new celestial bodies that brought him the greatest rewards of position, status, and authority. The telescope had been developed in the Netherlands in the first years of the seventeenth century. Galileo heard of this invention and imported a model. He worked out the optical principles and developed a more powerful version. In 1609 he demonstrated his marvel to the Venetian court, showing that his backers could see a returning merchant ship through the telescope two hours before someone searching with only the naked eye, thereby allowing a manipulation of the commodities market. Here was insider trading with a vengeance! The Venetian Senate was very impressed. They were willing to double his salary and give him a lifelong position, but in return all his future inventions would belong to the Senate and he could never ask for another pay increase.

Galileo had his eye on another prize—a position at the Medicis' court in Florence. The Medicis were arguably the most important patrons on the Italian peninsula, surpassed only by the court of the pope for power and prestige. They were possibly the richest family in Europe, with connections to the pope and business interests all over the Mediterranean and beyond. They could give Galileo the status and freedom he desired. Galileo began by teaching mathematics to the Grand Duke's son, Cosimo (descendant of the Cosimo who founded the Platonic Academy). He invented the proportional compass (which he manufactured and sold) and for a time taught practical mathematics like navigation. In 1609 he made his move.

A. Setting cannon inclination.

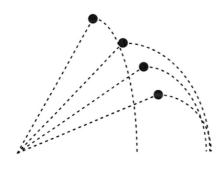

B. Various ranges by angle prior to Galileo.

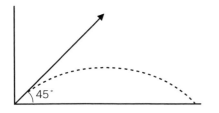

45°

C. Galileo's solution.

4.7 THE QUESTION OF CANNON RANGE

He turned his telescope on the skies and discovered that there were four moons circling Jupiter. This and other findings were all published in *Sidereus Nuntius* (*The Starry Messenger*) in 1610. He also discovered that the sun had rotating spots, that Venus had phases, and that the moon had craters and mountains. All this was highly controversial, since it showed the imperfection of the heavens, which went against Aristotelian supralunar perfection. He named the four moons of Jupiter the Medician stars, as a gift from a prospective client to a powerful patron. Cosimo, now Grand Duke, was delighted. After much negotiation, Galileo was given the position of Court Philosopher. This was a huge jump in status. But, of course, with status came risks. Galileo was now expected to take part in many intellectual wrangles as duels for the honour of his patron. Eventually, he moved from Florence to Rome and looked to the pope for patronage. These risks proved his downfall.

Johannes Kepler

Another astronomer whose career was equally influenced by these new patronage requirements was Johannes Kepler (1571–1630). Kepler was an anti-social, near-sighted man, descended from a family of misfits. Despite this, he became the Imperial Mathematician in the court of the Holy Roman Emperor Rudolph II, and thereby joined the practice of astronomy to the glory and wonder of this powerful court. Kepler is often called the first true Copernican (although several lesser-known sixteenth-century astronomers could share this title) because he whole-heartedly endorsed the heliocentric system. In the process of doing so he changed it to one that would have horrified Copernicus, since he destroyed the idea of the perfect circular motion of the heavens. He also attempted to join the physics of the heavens to a mathematical model of their motion. In other words, Kepler asked what the physical cause of the motions of the heavens was, rather than just mapping their course. His explanations were not taken up by other natural philosophers but showed astronomers that such questions were important.

Kepler was conceived on May 16, 1571, at 4:37 am and was born on December 27, 1571, at 2:30 pm, after a pregnancy lasting 224 days, 9 hours, and 53 minutes. We know this because Kepler cast his own horoscope, and these details were necessary to make accurate predictions. This demonstrates the importance of astrology to Kepler in particular and to early modern astronomers and society more generally, as well as the importance for Kepler of precision and mathematical accuracy.

Kepler had a very unhappy childhood. He grew up in a very poor Swabian Lutheran family with an abusive father and an unbalanced mother who was later tried as a witch. The high point of his young life was receiving a scholarship to the University of Tübingen, where he studied theology. When he finished his degree, he took a job as a mathematics teacher and astrologer in Graz. While teaching mathematics (to virtually empty classrooms—his pedagogic skills were low), he had a revelation that was to change his life. In a flash of insight, the structure of the universe was laid bare to him. He was circumscribing a triangle with a circle when he realized that the orbits of the planets might work this way. (See figure 4.8.)

From this Kepler developed three questions: Why were the planets spaced the way they were? Why did they move with particular regularities? Why were there just six planets? (The latter question marks him as a Copernican, since there are seven planets in the Ptolemaic scheme.) With his insight concerning the circumscribed triangle, he saw the answer to the first and last questions. He transformed his two-dimensional figure into a three-dimensional solid. Since there are only five regular solids in Euclidean geometry, the six planets fit perfectly with one solid between each orbit. This seemed to recreate the particular spacing of the planets. Kepler published this finding in *Mysterium Cosmographicum* (*The Mystery of the Universe*) in 1596. (See figure 4.9.) Later, in his *New Astronomy* (1609) and in the second edition of *Mystery*, he laid out the physical reason for the planets' motion in this particular configuration. He postulated that some sort of "magnetic" force emanated from the sun, in the centre, and was the cause of motion. That is, the sun was the prime mover, a concept that shows that Kepler was influenced by neo-Platonic ideas. Although there were many problems with this whole schema, it would provide Kepler with his life's project.

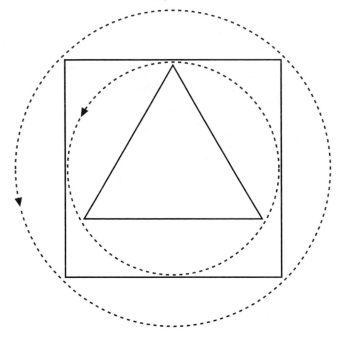

4.8 KEPLER'S ORBITS

Rotating the triangle and the square produces orbits.

Kepler recognized that in order to improve his model he needed better observations of planetary motion. He decided to go to the best observer in Europe and so became an assistant to Tycho Brahe. Kepler joined Tycho in Prague, where Tycho had recently become the Imperial Mathematician to Rudolph II. They had a very stormy relationship. Tycho insisted that Kepler work on the orbit of Mars, which he was not very happy to do. As it turned out, this was very fortunate, since Mars has the most irregular orbit of all the planets, and Kepler was forced to abandon the idea of a circular orbit in order to match observation to mathematical model. Kepler never did his own observations (he was far too near-sighted to see the stars and planets accurately), but he spent eight years calculating sheet after sheet of numbers. This was boring, repetitive, exacting work with little reward.

Tycho had hoped Kepler would prove the Tychonic system, but Kepler, as a Copernican, had other plans. After Tycho's death in 1601 Rudolph appointed Kepler Imperial Mathematician in his place. This gave Kepler status, although not much pay.

4.9 KEPLER'S NESTED GEOMETRIC SOLIDS

Based on Kepler's concept of the spacing of the planets in *Mysterium Cosmographicum* (1596).

He earned his living casting horoscopes but still had time to work on what is often seen as his greatest work, *Astronomia Nova* (1609) or, more fully, *A New Astronomy Based on Causation or a Physics of the Sky Derived from Investigations of the Motions of the Star Mars. Founded on the Observations of the Noble Tycho Brahe.* Kepler had worked on the orbit of Mars for eight years and had got his calculations to agree with the Copernican system to within eight minutes of arc. Although this is fairly accurate (Copernicus himself was only accurate to within 10 minutes), Kepler was sure that Tycho's observations were better than that. After a terrible struggle, he concluded that the orbit of the planet was an ellipse. Although this was not the first law of planetary motion he worked out, astronomers and historians came to

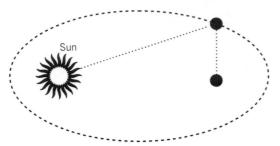

LAW 1: Planets move in elliptical orbits with the sun at one focus.

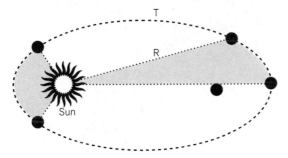

LAW 2: AREA LAW. The time to sweep out areas of equal size takes equal time.

$$(T_1/T_2)^2 : (R_1/R_2)^3$$

LAW 3: PERIOD LAW. The square of the period "T" (time to complete one orbit around the sun) of two planets is proportional to the cube of the distance "R" to the sun.

4.10 KEPLER'S THREE LAWS

call it Kepler's First Law because it underlay his other observations. He also postulated that the "magnetic" force of the sun, sweeping the planets around before it, operated in a mathematically consistent way and that a line from the sun to each planet swept out an equal area in an equal time (called the equal area law, or Kepler's Second Law). This meant that, when the planet was closer to the sun, it moved faster.

In 1618 Kepler published the third of his great books, *Harmonices Mundi* (*The Harmonies of the World*). In this work he argued that the planets, sweeping out their paths through the heavens, created harmonious music. It is perhaps telling that Kepler wrote this book that claimed to have discovered a grand scheme of harmony—in music, astronomy, and astrology— at the start of the Thirty Years' War, which necessitated his flight from Prague, and during the trial of his mother for witchcraft and the death of his daughter. As Kepler says in *Harmonices Mundi*, "In vain does the God of War growl, snarl, roar, and try to interrupt with bombards, trumpets, and his whole tarantantaran … let us despise the barbaric neighings which echo through these noble lands, and awaken our understanding and longing for the harmonies."[2]

In the midst of discovering the major third played by Saturn and the minor third played by Jupiter, Kepler also developed what we now call his Harmonic Law, or Third Law. In this law, developed by trial and error, he demonstrated the

2. Johannes Kepler, Dedication of the *Ephemerides* (1620) to Lord Napier, in *Harmonies of the World*, as quoted in Arthur Koestler, *The Sleepwalkers* (London: Penguin, 1959) 398.

mathematical relationship between the periodic time (time for a single revolution around the sun by a planet) and the distance from the sun, so that the farther away from the sun, the greater the periodic time. He found that the ratio of the period of the orbit squared (T^2) to the mean radius of the orbit cubed (R^3) is the same value (K or a constant) for all the planets.

Despite the amount of work Kepler did, his explanations for planetary motion had little impact on other astronomers of his day. As the Imperial Mathematician to Rudolph's court, he was an important representative of natural philosophy. His books were certainly taken seriously but, except for the *Rudolphine Tables*, seem to have been seldom read. In his time his work was regarded as difficult and even dangerous. His place in the history of science depends more on his relationship to later ideas than his effect on astronomy of the day. Historians have selected the three "Laws" that accord with more modern astronomical ideas, particularly as identified by Newton, but they were mixed in with dozens of other laws created by Kepler and now forgotten. Galileo, Kepler's contemporary, saw him as a dangerous person to know, and their correspondence was polite and unenthusiastic. Kepler was suspect: as a Protestant, as a rival court astronomer, and as someone known to travel close to "occult forces," both because of the witchcraft accusation levelled at his mother and because his physical explanation for the motion of the heavens relied on action at a distance. Action at a distance required things to interact without some material connection between the objects, and Kepler's speculation about a kind of magnetic force moving the planets was seen as a magical explanation.

Newton gave credit to Kepler for a number of ideas but later asserted that he got nothing from Kepler's work. On the other hand, Kepler shows us how astronomy worked in the sixteenth and early seventeenth centuries. His years of calculating demonstrate the importance of mathematics to the study of the universe, and his place at Rudolph's court reminds us of this new site of natural philosophical knowledge.

The Protestant Reformation and the Trial of Galileo

The courts allowed men of practical knowledge, sometimes skilled artisans and mathematical practitioners, to mingle with university-trained or self-taught natural philosophers. They brought together these different ideas and interests and in doing so created new questions and goals for natural knowledge. Most natural philosophers attached to princely courts gained their reputations both for intellectual acuity and for practical applications. For example, Kepler and John Dee cast horoscopes for Rudolf and Elizabeth respectively. Dee advised Elizabeth on the most propitious day for her coronation, as well as consulting with navigators searching for a northwest passage. Likewise, Galileo's activities as a courtier were both esoteric and applied. These men walked a fine line between theory and practice, since all three were interested in large philosophical systems and desired court patronage not simply for creating improved telescopes or new armillary spheres. But monarchs wanted results, and all investigators of the natural world with court connections were compelled on occasion to dance for their supper. So claims to utility and the search for topics interesting to those princely patrons changed the orientation of natural philosophy away from philosophical speculation toward how things worked.

One good reason for natural philosophers to avoid philosophical speculation or the more traditional career path of Church positions was the other huge upheaval of the sixteenth century, the Protestant Reformation. While protests against various perceived inadequacies of the Catholic Church had flared up in the fifteenth century, Martin Luther's decisive stance in 1517 against indulgences split the Catholic Church in two. Just as with natural philosophy, religion was affected by the printing press—the printing of those indulgences flooded the market and made the venality of the Church more obvious, while the pamphlets printed by Luther's supporters and detractors ensured that there was not a corner of Europe that didn't know about the conflict within a few years.

The Reformation changed the intellectual, social, and institutional worlds in which natural philosophers lived. No longer did the Catholic Church have a monopoly on truth, which was either wonderfully liberating or terrifying, depending on your religious position. There were new career possibilities and new places where a study of nature might be useful, such as merchants' houses, princely courts, and more secular private schools. While leaders on both sides of the religious divide called for a return to salvational concerns rather than secular ones, a window had been opened for alternative thinking and careers.

There has been much debate among historians as to the effect of the Reformation on science. Some have pointed to the flourishing of science in strongly Calvinist or at least Protestant areas as evidence of the support for science in Protestant attitudes. Others have pointed to the Catholic Church's treatment of Galileo to show the devastation caused by "superstition." The truth is that the impetus for people to investigate natural philosophy often was a way of removing themselves from sectarian strife, of finding a middle way of worshipping God through his works. The crisis of Galileo was nothing as clear as Catholics against science. A large part of Galileo's modern fame comes from his image as the "Defender of Science." However, he got into trouble not because he defied the Catholic Church but rather because he attempted, unsuccessfully, to reconcile science and religion and because his patronage choices proved too risky. All his life Galileo remained a staunch Catholic. He believed that his real enemies were not the Church authorities but "philosophers"—the Aristotelians who argued that only they had the right to make truth claims about the world.

In 1614, shortly after Galileo's astronomical discoveries, which had been hotly disputed, Galileo, and with him Copernicus, were attacked from the pulpit. Galileo, although sick, entered the fray. He wrote a letter explaining the division of knowledge between nature and scripture. When this letter fell into the wrong hands, he sent a longer version to Cardinal Bellarmine and went himself to Rome to explain the situation. Bellarmine was a humanist and moderately sympathetic to Galileo's situation. While some churchmen, especially Dominicans, believed that the motion of the Earth was unprovable, Bellarmine held it to be unproven. This was a softer position, although it is highly unlikely that Bellarmine believed that such a proof could be found.

Galileo decided to make his position clearer. In an extended version of this earlier letter, the "Letter to the Grand Duchess Christina," which was written to be circulated, Galileo claimed (following Augustine) that there must be a separation between science and religion in order to maintain the dignity of both. In an argument that stood Thomas Aquinas on his head, he argued that scripture can never be used to disprove something that has been proved by observation and right reasoning; rather, scripture must be reinterpreted to take this into account. He did not make this argument against the Church or Christianity. Instead, he was concerned that Catholic natural philosophers would lose status to Protestant ones and that the true wonders of God, as understood in his work, would not be observed and interpreted. Galileo quoted an early Church father, to very different effect: "That the intention of the Holy Ghost is to teach us how one goes to heaven, not how heaven goes."[3]

Galileo took this risky and public stand in part because of his loyalty to the Catholic Church and his desire for a strong natural philosophical community in Italy. Equally, this could be seen as a move to be noticed by the pope to whom he was looking for patronage. A client had to take risks to maintain client visibility if

....

3. Galileo Galilei, "Letter to the Grand Duchess Christina" (1615), in *Discoveries and Opinions of Galileo*, ed. Stillman Drake (New York: Doubleday Anchor Books, 1957), p. 186.

he hoped to be successful. So he went on the offensive. Hearing rumours that both his Letter and the works of Copernicus were about to be placed on the Index and therefore unavailable for good Catholics to read, Galileo went once again to Rome to seek an audience with the pope, Paul v. Instead, he had a meeting with Bellarmine, at which he was instructed to stop work on the Copernican theory. The judgement of the papal tribunal was that this theory was "foolish and absurd in philosophy," and an Interdict was produced in 1616, which told Galileo that he was no longer to hold or defend the Copernican theory.

In 1623 three new comets appeared in the heavens, and Galileo was drawn once again to astronomy. In the meantime Paul v had died, and in his place was a humanistic pope, Urban VIII. Galileo, thinking he had an ally in the papacy, visited Urban VIII in 1624, asking to be allowed to write about the Copernican system. He neglected to mention the earlier Interdict. Galileo left the audience believing he had received permission to write about it in a hypothetical manner. Ultimately, this turned out not to be the case.

In the 1620s Galileo began to develop his defence of Copernicanism, resulting in *The Dialogue Concerning the Two Chief World Systems* (1632). The dialogue form allowed him to present both sides of the argument (Ptolemaic and Copernican) without definitively choosing one, but since the character espousing the Ptolemaic system was named Simplicio, it was not hard to see Galileo's inclination. He used his theory of the tides as a proof of the motion of the Earth and, hence, of the Copernican doctrine. Although the theory was quite wrong-headed and convinced no one, it demonstrated to those reading the book that Galileo was indeed defending Copernicanism and so breaking the Interdict of 1616, which prohibited holding the view that the Copernican system was a proven fact. If that was not enough, the pope believed he had been personally betrayed by Galileo.

Galileo was called before the Roman Inquisition in 1632, and the trial took place in 1633, after he arrived in Rome. The trial revolved around whether Galileo had been ordered not to *teach* the Copernican system. Galileo said that wasn't part of the Interdict document he had received from the papal office but, rather, that he had been told he could not *hold* the system to be true, which was what all Catholics were enjoined to believe. The Inquisition stated that Galileo, and Galileo alone, had been told he could neither hold, teach, nor in any way defend the Copernican theory. This came from a document that was either a forgery or an unissued draft of a papal directive. Thus, this was not a trial of science *versus* religion but a matter of obedience to the Church. The Inquisition judged that Galileo had disobeyed, and they effectively silenced him. He was placed under house arrest for the rest of his life. He returned to his study of mechanics and wrote his most brilliant work, *The Two New Sciences*, also written as a dialogue. Since no Catholic was allowed to publish any of his work, the manuscript was smuggled out to the Protestant Netherlands. All his work, especially the two dialogues, became very popular and were translated into several languages.

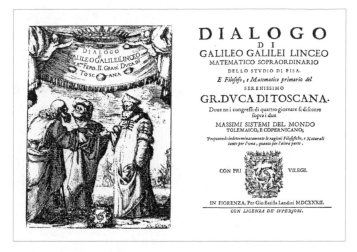

4.11 FRONTISPIECE FROM GALILEO'S *DIALOGUE CONCERNING THE TWO CHIEF WORLD SYSTEMS* (1632)

The silencing of Galileo and the shift of scientific work to Protestant areas of Europe have suggested to some historians that Protestantism was more conducive to science. This is problematic in general terms, since natural philosophers continued to flourish in France, particularly within the Jesuit order, while Protestant religious leaders were often far more antagonistic to the study of nature than Catholics. In terms of the pursuit of utility, however, where knowledge of nature was seen as useful for mercantile, empire-building nations, the Protestant regions were far more willing to pursue science as a study. The whole idea of the reasonableness and simplicity of nature, although not exclusive to Protestantism, was emphasized by them, and natural philosophers also looked for the simplest answer. Protestants emphasized the idea that knowledge should be useful, either for human betterment or salvation, and natural philosophers often directed their studies to topics that had utility (or claimed it). Protestants felt that God had given them the Earth to exploit to its full extent, and exploitation became an underlying ideology of science. Puritans and Calvinists believed in personal witnessing and experience; scientific methodology increasingly employed experiment. The idea that the individual could find his own way to God through private study was borne out in science. Theories of election and vocation led not only to the idea of the investigator of nature as purer and higher but even to the cult of the scientist. Protestants rejected Church traditions; the New Science rejected traditions of Aristotelian science. Finally, Calvinism and Puritanism, especially, appealed to the urban mercantile classes, those people interested in the questions of exploration, navigation, astronomy, and mathematics, which would be the breakthroughs of the New Science.

The arguments made about the pursuit of natural philosophy in the regions controlled by Protestants and those controlled by Catholics are less about the direct relationship of natural philosophy to religion and more about the intellectual space created by the conflict. While the Catholic hierarchy may have silenced Galileo, the study of nature still remained important. Luther vehemently rejected Copernicanism, but it did not follow that Protestants abandoned astronomy. If people could question the very nature of religious faith as the Protestants did, then intellectually no question seemed out of bounds, whether for Protestants or Catholics. For a small group of people, natural philosophy seemed to offer a "third way" of worshipping God in a world where secular authority was unreliable and religious authority was wracked by dissension and uncertainty. Nature was consistent, unlike the pronouncements of monarchs, popes, and priests.

Education and the Study of Nature

Education, the other principal institution important for natural philosophers, was rapidly changing in early modern Europe. Previously, education had been largely an ecclesiastical concern. Most schools were sponsored by the Church, and many schoolmasters were clerics. From the mid-fifteenth century on, secular interest in education began to rise, first in Italy and later throughout Europe. The goal of education ceased to be only a career in the Church; government offices, secretarial positions, and eventually gentry culture and patronage possibilities all provided new incentives for achieving a certain level of education. At the same time the Protestant Reformation produced a new impetus for education and literacy, both because Protestants argued for the importance of personal and vernacular Bible reading and because the Catholic Church responded, in part, through educational strategies. Thus, education became a *desideratum* for a wider sector of the population.

A significant minority were very well educated, and increasingly during the early modern period these well-educated men were in positions of social, political, and economic power. As well, some men and women were self-taught or continued their education on an informal basis throughout their lives. Because of this increasing market for educational currency, institutions such as the universities developed less formal curricula, designed for those not interested in a credentialed profession. Even the more traditional subjects such as medicine began to look toward greater applicability of their knowledge. Other kinds of academies sprang up all over Europe to cater to specialized learning. Self-help books became more and more popular, and educational entrepreneurs, both humanists and others such as mathematical practitioners, began to sell their educational wares through individual lessons and books. Thus, this early modern period witnessed a change in the status of education among the governing classes across Europe, especially in the north and west, and thus in the demand for both educated advisors and information itself. In this climate, the utility of the subjects studied became important.

Andreas Vesalius

Andreas Vesalius (1514–64) was trained and pursued his career in this new university structure, as well as being influenced by humanism. Born in Brussels, Vesalius was the son of an apothecary of the Emperor Charles v. In 1530 he attended the University of Louvain and then moved to Paris to pursue a medical degree. In 1537 he enrolled at the University of Padua, renowned for its medical school. Almost immediately he received his Doctor of Medicine Degree and became an anatomy lecturer, a rather low-status occupation. He caused a sensation by insisting on performing dissections himself. This was almost unheard of, as anatomy lecturers traditionally had read from Galen while their assistant pointed to the pertinent parts. Vesalius soon began travelling around Italy and the rest of Europe performing public dissections. He rapidly found problems with the traditional Galenic anatomy, since it did not correspond to what he was seeing. This observation was possible only because of his personal interaction

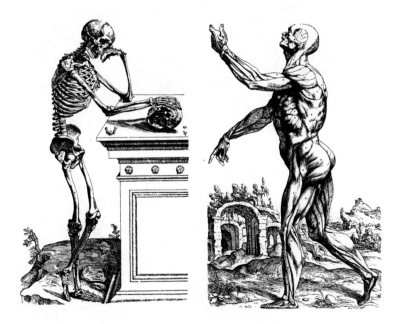

4.12 BONE AND MUSCLE MEN FROM VESALIUS, *DE HUMANI CORPORIS FABRICA* (1543)

ANDREAE VESALII
BRVXELLENSIS, INVI-
ctissimi CAROLI V. Imperatoris
medici, de Humani corporis
fabrica Libri septem.

CVM CAESAREAE
Maiest. Gallorum Regis, at Senatus Veneti gratia &
privilegio, ut in diplomate eorundem continetur.

4.13 FRONTISPIECE FOR VESALIUS'S *DE HUMANI CORPORIS FABRICA* (1543)

with the bodies. In 1543 he published the results of his disagreements with Galen, a new method, and a new philosophy in *De humani corporis fabrica* (*The Fabric of the Human Body*). Vesalius produced a beautiful book and in the process disproved a number of Galen's ideas.

Vesalius showed that the liver was not five-lobed but one mass, that men did not have one less rib and women one more, that nerves were not hollow, and that bones were a dynamic foundation of the human body. For the first time he pictured the muscles in a rational methodical way. Perhaps his greatest achievement, however, was his method. He began with humanism, since he compared alternate texts of Galen in order to find the purest and least corrupted. Once he had questioned the text, he turned back to its source, the human body. He then used observation, seeing personal experience as fundamental for the natural investigator. Rather than relying on authority, Vesalius prescribed first-hand dissection for all would-be anatomists.

There is, of course, an irony here, since Vesalius's book soon achieved the same level of authority he had derided in Galen!

Equally important, Vesalius's dissections were not done in a closed academic forum but in public. He helped to establish public demonstration and witnessing as an important part of natural science. As the frontispiece to *De fabrica* shows, the knowledge of the human body was gained because everyone saw personally, yet together, and all agreed on what they had seen. (See figure 4.13.) This idea of public demonstration as the creation of knowledge, of matters of fact, and the need to have a group of like-minded individuals to agree on closure, became a necessary ingredient to scientific practice and discourse in the seventeenth century and beyond.

Conclusion

The establishment of natural philosophy as an enterprise to be conducted in public, in the universities, the courts, the merchant halls, and the instrument makers' shops was an innovation of the Renaissance and early modern period. The rediscovery and printing of ancient knowledge had, ironically, allowed early modern scholars to claim that they were now developing new knowledge rather than conserving what existed. The changing social, political, and religious world gave these scholars new venues to investigate nature and new claims to the secular utility of their task. All natural philosophers of this period believed that to study nature was to study God's work and that this was a sacred task, but equally they believed that the point of this enterprise was firmly rooted in the present, with the goal of human betterment or personal advancement (and maybe both). In an era of great adventure and discovery, Western Europe, especially those countries on the Atlantic, began to believe in the possibility of boundless progress. Had they not already surpassed the ancients in exploration of the Earth with Columbus's voyages and of the heavens with Galileo's telescope? The courtly natural philosophers were men of creativity and action, not austere and academic theologians. This made all the difference to their attitudes and to the face of science.

Essay Questions

1. What influenced Copernicus to develop a new heliocentric system?

2. How did voyages of exploration influence scientific understanding of the world?

3. What role did patronage play in the development of science in the sixteenth and seventeenth centuries?

4. Why did Galileo come into conflict with the Roman Catholic Church?

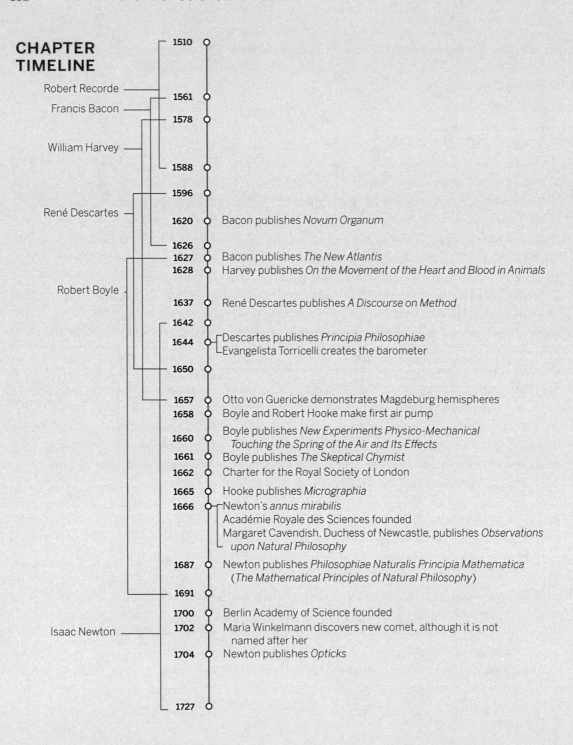

CHAPTER TIMELINE

Robert Recorde

Francis Bacon

William Harvey

René Descartes

Robert Boyle

Isaac Newton

1510	
1561	
1578	
1588	
1596	
1620	Bacon publishes *Novum Organum*
1626	
1627	Bacon publishes *The New Atlantis*
1628	Harvey publishes *On the Movement of the Heart and Blood in Animals*
1637	René Descartes publishes *A Discourse on Method*
1642	
1644	Descartes publishes *Principia Philosophiae*
	Evangelista Torricelli creates the barometer
1650	
1657	Otto von Guericke demonstrates Magdeburg hemispheres
1658	Boyle and Robert Hooke make first air pump
1660	Boyle publishes *New Experiments Physico-Mechanical Touching the Spring of the Air and Its Effects*
1661	Boyle publishes *The Skeptical Chymist*
1662	Charter for the Royal Society of London
1665	Hooke publishes *Micrographia*
1666	Newton's *annus mirabilis*
	Académie Royale des Sciences founded
	Margaret Cavendish, Duchess of Newcastle, publishes *Observations upon Natural Philosophy*
1687	Newton publishes *Philosophiae Naturalis Principia Mathematica* (*The Mathematical Principles of Natural Philosophy*)
1691	
1700	Berlin Academy of Science founded
1702	Maria Winkelmann discovers new comet, although it is not named after her
1704	Newton publishes *Opticks*
1727	

THE SCIENTIFIC REVOLUTION: CONTESTED TERRITORY

5

From 1543 to 1687 some of the giants of science lived, contemplated the natural world, and produced the underpinnings of modern science. Given the accomplishments of the period, it is no wonder that historians have agonized about the idea of an era of scientific revolution. Indeed, the twentieth-century discipline of the history of science began by focusing on the problem of the origin of modern science. The work of some of the discipline's founders concentrated on what this important transformation was and how it came to take place. In recent years, historians have begun to question whether or not such a revolution happened at all. Obviously, the answer depends on how revolution and science are defined: whether there was a gradual transformation of ideas, a gestalt switch, or a sociological innovation. We argue that there was a transformation in the investigation of the natural world, in which new ideas, methods, actors, aims, and ideologies vied with one another for a newly secularized role in the developing nation-states. This, indeed, was a scientific revolution.

The scientific revolution can be understood as a series of overlapping innovations, all important in the creation of modern science. First, natural philosophers took up the epistemological challenges of the ancients and developed a new methodology for uncovering the truth about the natural world. Second, in many different areas, but particularly in physics, astronomy, and mathematics, new theoretical models of the universe were developed. Further, those interested in the investigation of nature formed new institutions and organizations that began to perform the now largely secular tasks of evaluating scientific fact and determining who could be a natural philosopher. Finally, and perhaps most significantly, men interested in the underlying truths of nature developed a new ideology of utility and exploitation, a new structure for scientific practice, and a gentlemanly coterie of scientists who applied their social standards of behaviour to the ideology of modern science.

The New Scientific Method: Francis Bacon and René Descartes

The rediscoveries of ancient natural philosophers and the challenges to that ancient knowledge in the sixteenth century caused scholars in the sixteenth and seventeenth centuries to turn to the epistemological question of how to determine truth from falsity. Perhaps spurred on by the religious turmoil of the sixteenth century, philosophers began to ask the question still fundamental to us today: How can we know what is true? This led to the development of a new form of scientific inquiry—a new "scientific method"—and a new way of articulating this search for certainty. In England this methodology was most fully articulated in the writing of Sir Francis Bacon (1561–1626). Bacon, although not himself a natural philosopher, proposed a reform of natural philosophy in the *Novum Organum* (1620) and *The New Atlantis* (1627). This program of reform was part of a grander scheme to transform all knowledge, especially legal knowledge and moral philosophy. Bacon believed that all human knowledge was flawed because of the Idols that all men carried with them. The Idols were the prejudices and preconceived ideas through which human beings observed the world. Bacon felt that the only way for natural philosophers to disabuse themselves of these Idols was to look at small, discrete bits of nature. The only way to be certain one understood these small bits was to study them in a controlled setting, isolated from the larger (uncontrolled) environment. Using this assumption, he introduced what has come to be called the inductive method. He suggested that increments of information could be gathered by armies of investigators, put together in tabular form, and explained by an elite cadre of interpreters.

Bacon described this in a section of *The New Atlantis* known as "Solomon's House." His methodology, although it appeared to be more democratic than earlier scholastic methods, proposed a means of controlling truth and knowledge by a small elite group who determined what could be studied and what answers were acceptable. In this attitude, he was probably influenced by the fact that he was trained as a lawyer and that he spent much of his political career as an advisor to Elizabeth I and then as Lord Chancellor for James I. He was thus accustomed to the idea of testing evidence in the public venue of a court. As the person most concerned with treason and heresy, he had a distrust of free-thinking and believed that ideas should be controlled by those whose position as custodians of the peace and security of the commonwealth best assured their credibility. One aspect of Bacon's job as Lord Chancellor included overseeing the use of torture in an age when evidence obtained by torture was considered reliable. For Bacon, knowledge was power, and thus an understanding of nature was important precisely because of the practical applications such knowledge would have. In many ways Bacon was a courtly philosopher, so the rhetoric of utility so well employed by Galileo was also present in his work.

This methodology was challenged on the continent by René Descartes (1596–1650) and his followers, who preferred a deductive style based on skepticism. Descartes, like Bacon, came from an influential family and was trained as a lawyer. Unlike Bacon, he worked as a mathematics teacher and practitioner, rather than as a courtier, although in the end the temptation for patronage overcame the struggle involved in living by one's wits. In 1649 Descartes, at the age of 53 the most famous philosopher of his time, accepted the post of Court Philosopher to Queen Christina of Sweden. This was a lucrative post that, like the patronage of the Medicis for Galileo, offered both financial support and status in exchange for glorifying Christina's court and providing philosophical services. Unfortunately for Descartes, whose health was poor, Christina's idea of using his philosophical services was to have him call on her three times a week at five in the morning to instruct her. He was dead of pneumonia before the winter was over. The Swedes sent his body back to France but kept his head. This led to low-grade tension between France and Sweden for close to 200 years, until in 1809 the Swedish chemist Berzelius somehow managed to get Descartes's skull and return it to the French scientist Cuvier, who reunited it with the body.

In *A Discourse on Method* (1637) Descartes offered the first early modern alternative to Aristotle's epistemological system: his method of skepticism. He began by doubting everything, peeling away all layers of knowledge until he came to the one thing he knew was true: that as a thinking, doubting being, he must exist in order to think the doubting thought. He encapsulated this idea in the famous declaration *Cogito ergo sum: I think, therefore I am.* From this starting point he developed through deduction from first principles a series of universal truths that he knew to be self-evident. This deductive method owed its origins to geometric proof, which starts with a small set of sure premises or axioms and proceeds to more complex conditions. Interestingly, although Descartes used this mathematical model and developed new mathematical methodology, most of his scientific theories were explicitly non-mathematical. Also, he was not interested in using experimentation as a means to discover knowledge about nature. Since the senses could be fooled, right reasoning was a much more reliable arbiter in natural philosophical debate than any crude experiment or demonstration might be.

Both Bacon and Descartes attempted to find ways of reasoning that would produce certain knowledge in an age where certainty was giving way to probability. Bacon answered the question of how we can know what is true in a careful, conservative way, involving a hierarchy of knowledge made to appear as a democratic republic of scholars. Descartes answered in an individualistic anti-communal way, which gave more power to individual thinkers but did not, in the final analysis, create a community of scholars.

Mathematics as the Language of Natural Philosophy

One result of this new methodology with its debate about reliable and sure knowledge was the gradual ascendancy of mathematics as the language of natural philosophy. Galileo had been convinced that God created the world in number, measure, and weight, and many other scholars interested in nature echoed this sentiment. For Aristotle, weighing or measuring a substance did not tell you anything interesting about it, but for those investigating nature in the sixteenth and seventeenth centuries knowing how heavy something was or how fast something went was surer knowledge than searching for final causes. They claimed that they only measured and observed rather than imposing underlying hypotheses they sought to prove. Sir Isaac Newton, for example, famously said *Hypotheses non fingo: I feign no hypotheses.* Following Galileo and Newton, natural philosophers increasingly looked for certainty through measurement rather than the analysis of cause.

Significant mathematical developments included the rediscovery of algebra and the development of the calculus; in addition, new and easier notation systems were devised. For example, Descartes instituted the use of a, b, and c for known variables and x, y, and z for unknowns. Mathematicians used these systems as they vied with one another for solutions to increasingly complex algebraic equations.

The creation of analytic geometry, primarily by Descartes but also by François Viète (1540–1603) and Pierre de Fermat (1601–65), placed a powerful tool in the hands of mathematicians and natural philosophers. By combining geometry and algebra, it became possible to transform geometric objects into equations and vice versa. This also opened the door to mathematizing nature, as everything from the trajectory of a cannonball to the shape of a leaf could be turned into a mathematical expression.

In turn, mathematicians began to look for new ways to measure areas under curves and to describe dynamic situations. The result was the invention of the calculus, independently developed by Isaac Newton in England and Gottfried Wilhelm Leibniz (1646–1716) on the continent. The calculus added infinitely small areas together under a curve or described the shape of the curve in formulaic terms. This allowed natural philosophers to accurately describe dynamic situations such as velocity and the motion of acceleration, something not possible with the older geometric and algebraic systems. Some mathematicians were concerned with the philosophical implications of the calculus, since it produced finite answers from the addition of infinite quantities, and the infinitely small and the infinitely large could, paradoxically, be equal. However, in an age that increasingly looked to the practical applications and utility of mathematics, the calculus proved to be an extremely fruitful device and was quickly taken up by the scientific community.

Leibniz was an influential German polymath, trained as a lawyer and employed most of his life by several German princes, especially the three dukes of Hanover. Unfortunately, his relationship with Duke Georg Ludwig (1660–1727) deteriorated, and when Ludwig became King George I of England, Leibniz was forbidden to enter the country. He was very critical of both Descartes's and Newton's work and became embroiled in an acrimonious dispute with Newton over who had invented the calculus first. This dispute, carried on by Leibniz in Germany and Samuel Clarke (1675–1729) in England, was one of the most famous philosophical disputes of the seventeenth and eighteenth centuries, as it ranged over issues of Newtonian natural philosophy and theology. In the end the participants died, and the matter was never resolved. It seems clear now that Newton and Leibniz did develop the calculus independently, with completely different notation and mathematical bases (Newton's was geometrical, while Leibniz's was analytical). In a sense Leibniz won, since the notation that has been used from the eighteenth century to the present is his rather than Newton's.

Mathematical Practitioners

One of the reasons mathematics became such a powerful tool in seventeenth-century natural philosophy was the presence of a new category of scientifically inclined men: the mathematical practitioners. Mathematics had been a quite separate area of investigation, and those interested in its issues had usually tied their studies to practical applications such as artillery, fortification, navigation, and surveying. In the early modern period, these mathematical practitioners provided the necessary impetus in the transformation of nature studies to include measurement, experiment, and utility. Their growing importance was the result of changing economic structures, developing technologies, and new politicized intellectual spaces such as courts, thus relating changes in science to the development of mercantilism and the nation-state. Mathematical practitioners claimed the utility of their knowledge, a rhetorical move that encouraged those seeking such information to regard it as useful.

Mathematical practitioners professed an expertise in a variety of areas. For example, Galileo's early work on physics and the telescope were successful attempts to gain patronage by using mathematics. Descartes advertised his abilities to teach mathematics and physics. Simon Stevin (1548–1620) claimed the status of a mathematical practitioner, including an expertise in navigation and surveying. William Gilbert (1544–1603) argued that his larger philosophical arguments about the magnetic composition of the Earth had practical applications for navigation. Leibniz used his mathematical power to act as an advisor on engineering projects for the dukes of Hanover. As well, many practitioners, including Thomas Hood (fl. 1582–98) and Edward Wright (1558–1615), demonstrated an interest in mapping and navigation.

This new interest in mathematics and in quantifying the behaviour of the world sparked interest in probability. Mathematicians did not believe the world was capricious but that our incomplete knowledge of it limited our comprehension. The introduction of a mathematical evaluation of probability was a step toward understanding complex systems in which not all the determining factors could be known with certainty. Blaise Pascal (1623–62), Pierre de Fermat, and Christiaan Huygens (1629–95) all investigated the mathematical basis of prediction of games of chance, which were popular pastimes in the seventeenth century. Pascal's interest in the geometry of chance had wider implications than gambling at cards and dice, since it led him to develop his probabilistic argument for belief in God, now known as Pascal's Wager. He concluded that, although one cannot know with complete certainty if God exists, by using four possible conditions, one's best probable outcome would result from belief in God. If God did not exist, one lost nothing by believing in Him, but if He did exist, and one believed, one would be saved. Conversely, one lost a great deal by not believing in God if He did exist, while gaining nothing if He didn't exist. (See figure 5.1.)

5.1 PASCAL'S WAGER

	Do Not Believe in God	Do Believe in God
GOD DOES NOT EXIST	Nothing gained or lost	Nothing gained or lost
GOD DOES EXIST	Damnation	Salvation

By the end of the century Jacob Bernoulli (1654–1705) had codified the mathematics of probability, arguing that mathematics gave us the greatest certainty possible in an uncertain world. The concept of probability was not well accepted in physics, however, where Newton's universal laws seemed to provide certain, rather than probabilistic, answers. The shift of the foundation of physics from certainty to probability was one of the most traumatic transitions in modern science, but it would not happen for almost 200 years.

New Models of the Universe

All these new attempts to find a path to certain knowledge were crucial, since natural philosophers were making some radical suggestions about the make-up of the cosmos. The scientific revolution is most clearly identified with the development of a heliocentric model of the universe. This began with Copernicus, who claimed that the Earth revolved around the sun and who developed a mathematical model to explain the movement of the planets. Natural philosophers had been slow to accept Copernicus's theory because it lacked a proper physical justification, such as Aristotle had provided through the concept of natural motion for his cosmological schema. Thus, the so-called Copernican Revolution (historians love to label innovations as revolutions) was incomplete until Sir Isaac Newton (1642–1727), the great English astronomer and mathematician, devised a mathematical model of motion that explained heavenly and earthly movement in a single physical system based on his concept of universal gravitation.

Isaac Newton: The Great Polymath

Isaac Newton was born on Christmas Day, 1642, coincidentally the same year that Galileo died. His father died before he was born, and his early life was not happy, spent in conflict with a stepfather he disliked and a mother who expected him to

During the sixteenth century, many mathematical practitioners made their living by giving lectures and instruction, selling instruments, and performing a variety of mathematically based activities such as surveying or casting horoscopes. The booming metropolis of London provided an excellent marketplace for these skilled men, whose work demonstrated the interconnection of natural philosophy, mathematics, and the mercantile society of the time.

The first mathematical practitioner in London was Robert Recorde (1510–88). Trained at Oxford, he was largely responsible for introducing arithmetic and mathematics to a wider audience in England and for re-establishing a mathematical language and discipline into English scholarship. Recorde wrote a number of foundational books on mathematics in English. He was commissioned by the Muscovy Company (a merchant company that traded in northern Europe and was seeking the Northeast Passage to Cathay) to give lectures and to write a series of books on geometry, spherical geometry, astronomy, and navigation for use by their navigators. At the same time, he introduced Euclidian mathematics and algebra to an English audience, allowing natural philosophers to use mathematics in ways they had not done before.

Recorde was soon followed by other mathematical lecturers in London, sponsored by guilds, companies, or the City of London itself. Thomas Hood (1556?–1620), for example, was the first mathematics

run the family farm. Newton had no aptitude for farming, and his mother despaired of finding him a livelihood. Fortunately, the local vicar noticed his scholarly potential and helped procure him a scholarship to Trinity College, Cambridge. Newton was a relatively undistinguished scholar, except for mathematics; he taught himself geometry from Descartes's works and algebra from Viète's. In due course, he was made a Fellow of Trinity College in 1664 and was appointed Lucasian Professor of Mathematics in 1669. The latter was a prestigious position, although he often lectured to empty rooms. His teacher, Isaac Barrow (1630–77), had to pull strings in order to get him this position, since Newton did not believe in the special divinity of Christ or in the Trinity. This made him a potential heretic, since he would not take the required oath of uniformity to the Church of England, which was normally required for any high academic or government position.

lecturer paid by the City of London. In 1588, Hood petitioned William Cecil, Lord Burghley, to support a mathematics lectureship in London, to educate the "Capitanes of the trained bandes in the Citie of London."* Hood identified himself on the title pages of all his books until 1596 as Mathematical Lecturer to the City of London, sometimes advising interested readers to come to his house in Abchurch Lane for further instruction, or to buy his instruments. His books explained the use of mathematical instruments such as globes, the cross staff, and the sector, suggesting that his lectures and personal instruction would have emphasized this sort of instrumental mathematical knowledge and understanding. He was also known to have cast horoscopes, another way to make money.

Those who gave and attended these mathematical lectures had some expectation that they would be able to buy, sell, and use the instruments that were being discussed there. It is no surprise, therefore, that just as these lectures were being presented to the London community, mathematical instrument makers were beginning to ply their trade in increasing numbers in late-sixteenth-century London. Thomas Gemini, a goldsmith, was probably the first English instrument maker starting in the 1550s, followed by Humphry Cole in the 1580s. After that, many men set up shop and sold navigational instruments, surveying instruments, maps, globes, and astrolabes, as well as providing the informal instruction for those who were interested in using them.

By 1610, there was a strong practical mathematical community living in London. A number of mathematical lectures had been sponsored, attended by a variety of audiences. Books and individual lessons explaining the use of mathematics and mathematical instruments had been produced, all leading to an increasing number of men trained in and sensitive to mathematical tools and explanations. A variety of men met in the instrument shops and at the mathematical lectures—gentlemen, scholars, merchants, and navigators. Mathematics was becoming both a language of commerce and of natural philosophy.

....
* British Library manuscript, Landsdowne 101, f. 56.

While Newton would not compromise his religious beliefs, he kept his views very private for his entire life.

In 1665 the Great Plague returned to England. It swept through Cambridge, and Newton was forced to return to his mother's home. The enforced isolation allowed him the opportunity to put together a number of ideas he had been developing through intensive reading during the past four years. Although it is unlikely that an apple really fell on his head, during the following year, his *annus mirabilis* (miraculous year), he worked out theories about gravitation, physics, and astronomy. As if that were not enough, he also created the calculus and began his investigations into optics and theories of light.

Although Newton studied planetary motion during that year, he did not publish his results until 22 years later. He was dissatisfied with his mathematical results to the question of why the orbits of the planets were circular, or nearly circular, and so put them away for a time in order to concentrate on alchemy and theology, including his long-standing interest in the books of Revelation and Daniel. He spent many hours over the next 13 years reading commentaries on scripture, studying the Bible, and constructing his personal theology. This theology was complex, most closely resembling an extreme form of Unitarianism or Arianism, which argued that Christ was not divine but the highest of God's created beings. Over his lifetime, Newton spent more time studying theology than any other subject.

In 1679 Robert Hooke (1635–1703), Corresponding Secretary of the Royal Society of London, wrote to Newton to find out what he was doing. It was Hooke's job to act as a kind of intellectual pen pal, communicating with members of the Royal Society and putting people with similar interests in touch. Newton and Hooke had a strained relationship, however, since Hooke had criticized Newton's earlier optical work. What, Hooke asked in 1679, would be the path of a body (a rock, for example) released from a high tower down to a rotating earth? Newton replied that he was not presently engaged in natural philosophical investigations, but he suggested that it would spiral east to the centre of the earth because the angular velocity from the top of the tower was greater than on the earth's surface. Hooke argued that this was not so: the path would be the result of a horizontal linear motion at constant speed combined with an attractive force toward the centre and varied inversely with the square of the distance between the body and the earth. This caused Newton to wonder whether his original question in 1666 as to why planetary orbits were circular was misdirected.

Newton set to work on mathematical models of elliptical orbits. In 1684 his friend Edmund Halley (c. 1656–1743, and the discoverer of the comet that bears his name) rode out to Cambridge to try to get Newton to communicate some of his mathematical work to the Royal Society, a difficult task since Newton was very secretive and refused to publish anything. Halley said that he and his friends, the architect and natural philosopher Christopher Wren (1632–1723) and Hooke, had wondered what motion an orbiting body would traverse if attracted to a central body by a force that varied inversely as the square of the distance between them. Newton replied that in 1679 he had proven to his own satisfaction that it would be an ellipse, although he did not have the demonstration available. Halley was deeply impressed, seeing this as a breakthrough in mathematical astronomy, and urged Newton to publish. He even promised to finance the publication. Newton agreed and in an astonishingly short period produced the book that changed astronomy and physics. Published in 1687, *Philosophiae Naturalis Principia Mathematica* (*The Mathematical Principles of Natural Philosophy*, usually known simply as the *Principia*) laid out Newton's theory of universal gravitation and established a mathematical and mechanical model for the motion of the whole universe. In this work he used the mathematical models of Galilean physics and combined them with the planetary models of Copernicus and Kepler.

Newton had produced a universal physics, a Grand Synthesis, which finally allowed astronomers to move away from Aristotle with confidence. He had produced a real model of the universe rather than merely a mathematical set of calculations, such as Copernicus had put forward.

In the *Principia* Newton defined force for the first time, adding it to matter and motion as the third essential quality of the universe. Force, he said, was necessary to compel a change of motion. Without force, a mass (another term he originated) would continue at rest or in rectilinear motion, due to inertia. This is often called Newton's First Law. His Second Law showed the way to measure force mathematically, now expressed as $f = ma$, and his Third Law asserted that for every action there is an equal and opposite reaction. In order to understand the motion of the bodies in the universe, Newton insisted that they operate in absolute time and space. In the Newtonian system, absolute time and space were independent and unchanging aspects of reality. Absolute time was uniform, progressing at the same rate of change, everywhere in the universe and at all past and future moments. People could not perceive absolute time directly, but could infer it mathematically by observing the motion of the planets and stars. Absolute space was the unmoving and constant dimension of the universe. This meant that all observers saw the motion of the stars and planets (and everything else) within absolute space, and thus they should all see the same motion as measured against the unchanging dimension of absolute space. The implications of this were profound. It meant that Newtonian physics was truly universal and someone measuring the universe from a distant star would get the same results as we would.

Newton took the concept of centrifugal force (the tendency of an object to fly away from a circular path) and stood it on its head. In his early work in 1666, he had followed the older view of circular motion, which sought to quantify and explain this tendency to fly away. But, in a major insight, the *Principia* argued for a centripetal force instead, a force that pulls objects into the centre. This is gravity. Thus, the moon's motion is composed of two parts: first, inertial motion carries it in a straight line; second, gravity constantly forces it to fall toward the Earth. The balance of these two forces holds the moon in orbit. (See figure 5.2.)

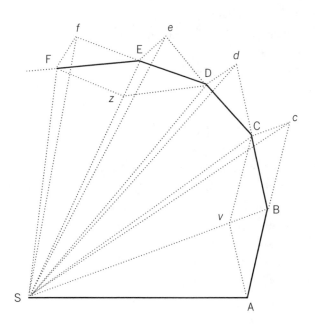

5.2 NEWTON EXPLAINS THE MOON'S MOTION

From Newton, *Principia Mathematica*, Proposition I.
At each point, A, B, C, etc., the moon's motion seeks
to move in a rectilinear line away from S, but is drawn
back by centripetal force (gravity) toward S.

Newton set up his argument in order to disprove Descartes's theory of vortices, which Descartes had articulated in *Principia Philosophiae* (1644). Descartes had argued that the universe was like a machine, a concept called "the Mechanical Philosophy." Except for those parts filled with coarse matter (such as the Earth), the universe was a plenum, or a space filled with an element called ether. The planets moved in a sort of whirlpool of ether that carried them along in their orbits. (See figure 5.3.)

The *Principia* was in many ways an attack on Descartes's book (the title reflected Descartes's), which is why Book Two concentrated on the study of how fluids work and how things move through them. This exploration of fluid mechanics may confuse modern readers, since it seems to have little to do with forces, gravity, or the motion of the planets. Newton's objective was not just to present his own system but to discredit Descartes's by showing that the vortex model was critically flawed.

The most far-reaching and long-lasting achievement of the *Principia* was how Newton found a way to tie together through the concept of universal gravitation the orbits of the planets and the satellites (including the moon and comets), Galileo's law of falling bodies, the fixation of objects to the Earth, and the tides. He proved that an apple falling from a tree obeyed the same laws as the moon orbiting the Earth. He had a law that applied to *everything*—the moon, the satellites of Jupiter and Saturn, the Earth, rocks, and the distant stars.

The sheer power of Newton's universal laws suggested to scholars in many different fields that there should be similar laws governing human interaction as well. Philosophers from many other areas, including economics and political philosophy, searched throughout the eighteenth century for such laws and argued that any society that failed to follow the universal laws was doomed to failure.

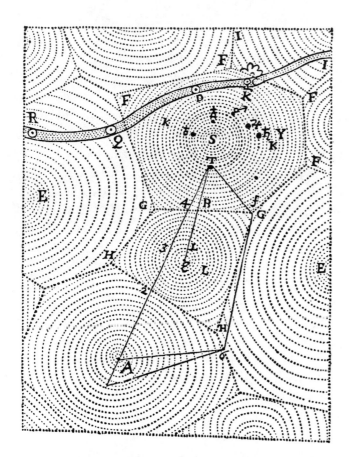

5.3 DESCARTES'S VORTEX COSMOLOGY
Points S, E, A, and Ɛ are centres of vortices. By the churning of the corpuscles within each vortex, the centre is self-illuminating, and thus is a star. Around S (the sun) move the planets, swept along by the vortex (T is the earth or terra). The stripe at the top indicates a comet's path. From Descartes's *The World or a Treatise on Light* (1664).

Ironically, the *Principia* was not a popular or financial success. The Royal Society, the pre-eminent institutional home of English science, refused to sponsor it, since their previous publishing venture, Francis Willoughby's *History of Fishes*, had been a financial disaster. They were unable to sell more than a handful of copies of this expensive illustrated book and were forced to give it to their employees, especially Robert Hooke, in lieu of salary! The members of the Royal Society, perhaps accurately, saw the *Principia* as a book with a limited readership and so forced Halley to fund it himself, which he did through the sale of subscriptions. This story reminds us that, as revolutionary as we might now consider Newton's work, at the time it held the interest of a limited audience because so few people could understand his mathematical arguments.

Newton and Alchemy

Because Newton considered his studies esoteric, accessible only to a select few, he was not overly concerned about reaching a wide audience. One aspect of his work that has been overshadowed by his physics and mathematics and so has faded from modern consciousness was his great interest in alchemy. Both Newton and Robert Boyle were involved in alchemical investigations, looking both for the Philosopher's Stone and the basic structure and functioning of nature. Although "multipliers," or alchemists interested only in gold, were looked upon as crass, many respectable gentlemen studied alchemy. Both Newton's and Boyle's investigations came from the belief in God's ordering of the universe and were a search for the active principles that animated nature, another attack on mechanical philosophy. Since Newton was interested in how the universe worked on a microscopic as well as a macroscopic level, he investigated what the prime matter of the universe was. He read both ancient and modern authors copiously and performed experiments of long duration. Because of his expertise in this area, in later years he was appointed Master of the Mint, in which position he assessed those who claimed to have produced gold. At his death, a number of people interested in alchemy were anxious to gain access to his extensive library of alchemical works.

In spite of his lack of bestseller status and his more arcane studies, Newton was the most famous scientist of his generation. He received many important marks of honour indicative of his high status, including his position as Warden and then Master of the Mint and election as President of the Royal Society. His opulent state funeral in 1727 demonstrated his exalted position, and reports of the event were transmitted around Europe, with Voltaire (who was deeply influenced by Newton's natural philosophy) providing one of the eye-witness accounts. Voltaire was extremely impressed that a natural philosopher, and one with heterodox religious views at that, was buried with such pomp and circumstance. Here was a country that understood the importance of its intellectuals!

Mechanical Philosophy

At the same time that Newton was introducing a mathematical basis for motion throughout the universe, other natural philosophers were searching for a more concrete model. Some, looking to the new instruments and machines being developed around them, argued that the world itself was a sort of machine. As men constructed more and more precision mathematical instruments, especially for navigation, astronomy, surveying, and timekeeping, and more complex machines, this suggested to them that God might have created the most complex machine of all. This conception of the world operating in a mechanical fashion became known as the Mechanical Philosophy or, alternatively, the atomical or corpuscular philosophy. It was first developed by René Descartes and Pierre Gassendi (1592–1655). Descartes especially extended the mechanical philosophy to include living things, arguing in his *Treatise on Man* (1662) that human physiology operated just like a machine. Corpuscular philosophy was later taken up by the Englishmen Robert Boyle and Thomas Hobbes, the latter of whom used it in his political theories. Basically, it was an interpretation of the world as a machine, either as a clock (implying order) or an engine (showing the power of nature). If the universe was a machine, then God was the Great Engineer, or Clockmaker. At first, Gassendi devised this interpretation of nature to give God a role as a transcendent, rather than immanent, being. That is, Gassendi argued that God could exist outside the material universe because He had set a mechanical structure in place. There was no need for God to exist within and tinker with an imperfect world. Corpuscular philosophy was eventually accused of being atheistic because, if the universe were a perfect clock, it would never stop, and there would be no need for God. The claim that natural philosophers were removing God from the universe was unfair but widespread, affecting Newton as well as Gassendi. It recurred in many forms and as a charge against many philosophers and scientists over the next centuries.

Mechanical philosophy was rooted in ancient theories of atomism, held by Epicurus, Democritus, and Lucretius, which had been rediscovered by the humanists. These ancient thinkers had posited that the world was composed of infinitely small simple particles. For the ancient Greeks, this had shown the eternity and total materiality of the world, an aspect of the theory that Descartes and especially Gassendi (a Catholic priest) set out to change. They argued that although it appeared that the world had existed forever, this could not be true, which meant that its creation was of a different kind than its operation and was, therefore, known by God, but unknowable through natural philosophy. A material universe could not be used as a proof for or against the existence of God, since God's existence was a metaphysical, not physical, question.

Mechanical philosophers reduced matter to its simplest parts, atoms, just as Descartes had stripped away ideas through skepticism. These atoms had only two qualities: extension and motion. Since extension was a definition of matter—that is, that matter must take up space and all space must be matter—a vacuum was not possible in this philosophy. Therefore, the universe was filled with a plenum of particles. All force-at-a-distance was actually motion through the plenum, which explained magnetism and the motion of the planets. It was this aspect of Descartes's theory that Newton had attacked so forcefully in the *Principia*. And the question of whether or not a vacuum could exist in nature was soon taken up in Robert Boyle's experimental program. It is a great historical irony that the close examination of the nature of matter, meant to prove mechanical philosophy, refuted the very theory that the universe was full of matter.

The Use of Experiments as Proof: William Harvey and Robert Boyle

Experimentation, as a source of sure knowledge about nature, was new to the early modern period. Aristotle, of course, had argued that forcing nature into unnatural situations would tell us nothing about how it really behaved. This attitude began to change in the sixteenth century, partly because of new attitudes toward certainty and man's power over nature, and partly because skilled instrument makers were able to create precise philosophical instruments. Francis Bacon, who as Lord Chancellor had overseen the torturing of traitors, believed that human beings, when subjected to extreme pain, would be forced to tell the truth. Likewise, putting nature on trial, including torturing nature through experiments, would force her to reveal her secrets. While the reliability of truth claims as a result of experimentation was under constant scrutiny in this period, it is fair to say that one of the significant changes to the study of nature in this period of scientific revolution was the increased use of, and reliance on, experimentation.

Probably the first extended discussion of experimentation came from the study of human anatomy in the work of William Harvey (1578–1657). Following the success of Vesalius in the sixteenth century, scholars developed a keen interest in the structure and function of living things. For example, Girolamo Fabrici (c. 1533–1619) examined the structure of veins and in 1603 found the existence of valves at particular intervals. Harvey, using keen observation, some well-thought-out experiments, and a belief in the similar structure of all animals including humans (what was later called comparative anatomy), developed a theory of the circulation of the blood that proved very influential in the years that followed.

Harvey received his medical training at Padua, where Vesalius had taught, and returned to London in 1602 to work first as a physician at St. Bartholomew's

Hospital and eventually as Royal Physician to Charles I. While in these positions he conducted a series of experiments on the blood in animals. This resulted in the publication of *On the Movement of the Heart and Blood in Animals* (1628), in which he demonstrated that the blood in animals and humans was pumped out by the heart, circulated through the entire body, and returned to the heart. He proved this through a series of elegant experimental demonstrations, some involving vivisection of animals and some less invasive demonstrations with humans. In his preface Harvey drew the political parallel between this circulation and the movement of citizens around their king, indicating his close affiliation with the Royalist side of the civil war soon to erupt in England.

One of the clearest experiments that Harvey performed was his proof that there were no passages through the septum of the heart. From Greek times through Galen and Vesalius, anatomists had argued that blood must pass through the wall separating the two ventricles or large chambers of the heart. While this explanation satisfied some aspects of the assumed purpose and path of blood, problems persisted. The first was the question of volume, since the old system required the liver to produce a constant supply of blood that was completely consumed by the rest of the body. This seemed to Harvey to be out of all proportion to the amount of matter that went into the body. The second problem was purely anatomical. While Galen had said there were pores in the septum, Vesalius, who worked with human hearts rather than the cow and pig hearts that Galen had used, could find no such passages. Rather than contradict Galen completely, Vesalius argued for a permeable septum with either sponge-like tissue or pores too small to see.

Harvey reasoned that if blood were circulated, a far smaller volume would be needed; it was simpler to conceive of a re-used supply of blood than a constantly created and consumed supply. At the centre of the system was the heart, working as a pump, but to demonstrate the two-part circulation (heart-to-lungs and back, heart-to-body and back) he had to show that blood did not move from the venous to the arterial system by way of the chambers of the heart. In a later experiment he demonstrated this, using a cow heart and a bladder of water. He tied off the passages to the heart so that he could squeeze water into one ventricle and see if it passed into the other. When it did not, he had the first proof that the blood did not pass through the septum. He then tied or untied the constraints and used the bladder of water to show that the blood must pass in sequence from right atrium to right ventricle and out to the lungs, and then through the pulmonary vein to the left atrium and left ventricle and out to the body through the aorta. While it was still unclear how the blood got from the arteries through the tissues of the body and back into the veins (and would remain unclear until microscopy developed far enough to see the microvessels at cell level), Harvey's work better explained the evidence than the older system. (See figure 5.4.)

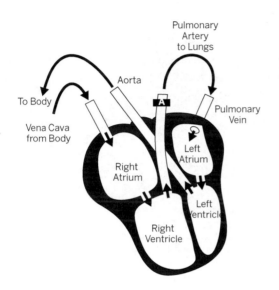

5.4 HARVEY'S MODEL OF THE HEART
Harvey closed off the artery at "A" and pumped water through the vena cava to demonstrate that no fluid passed from the right ventricle to the left ventricle.

Harvey also used careful observation and some experimentation in his embryological studies. Fabrici, in his book *On the Formation of the Egg and the Chick* (1621), had observed that in viviparous generation the embryo was created by a union of semen and blood from the parents. Harvey followed this work with a close examination of the development of ova. He examined fertilized eggs from their unformed state to birth, tracing even more closely the stages of growth. He published this work in *Exercises Concerning the Generation of Animals* (1657). This spurred further work by Marcello Malpighi (1628–94), who introduced the use of microscopic observations of ova development in 1672.

The work of Harvey and Malpighi, showing the power of observation and experiments, accorded with the inclination among many natural philosophers to use instruments and demonstrations to isolate phenomena and break down investigations into smaller components. Perhaps the man most responsible for the elevation of instrumental investigation during the scientific revolution was Robert Boyle (1627–91). Boyle, the son of an Irish noble family, came to Cambridge during the English Civil War and became a key player in the creation of the Royal Society after it. He joined the elite of London society from the 1660s onwards, living with his sister, Lady Ranelagh, in her house in Pall Mall, and entertaining royal and scholarly guests alike at his laboratory there. While at Cambridge, he set up his own alchemical laboratory to investigate the underlying make-up of matter. He denounced old-fashioned alchemical investigations and, in *The Skeptical Chymist* (1661), laid a foundation for the new study of chemistry.

Boyle, with Robert Hooke as his assistant, investigated airs, using a newly devised air pump. He employed instrument makers to attach a large, carefully blown glass globe to a pump in order to evacuate the air from the globe. In this, he

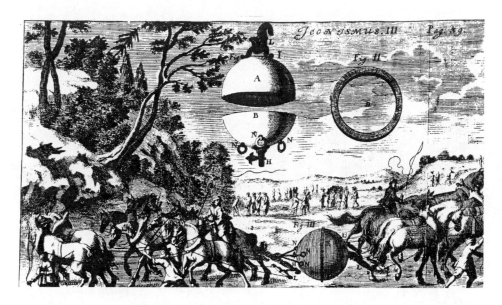

5.5 ILLUSTRATION OF VON GUERICKE'S MAGDEBURG EXPERIMENT

followed the lead of Otto von Guericke (1602–86), who had performed his own air experiments in the 1640s. Most famously, in 1657, von Guericke demonstrated that two teams of horses hitched to joined hemispheres of copper (the so-called Magdeburg hemispheres) from which the air had been evacuated could not pull them apart, since the weight of the air outside the spheres was so much greater than that on the inside. (See figure 5.5.)

Boyle and Hooke built their first air pump in 1658 and performed a number of experiments. (See figure 5.6.)They demonstrated that air had weight, that a vacuum could exist, and that some component of air was necessary for respiration and combustion. Their results were published in 1660 as *New Experiments Physico-Mechanical Touching the Spring of the Air and Its Effects.* Boyle used his air pump for less spectacular demonstrations than von Guericke, for example, placing small animals in the glass sphere, removing the air, and watching them perish, or alternately placing candles therein and watching the flame extinguish. (See plate 4 for a later depiction of these events.) From these experiments he concluded that there was something in the air that supported both life and combustion. His work was thus connected to Harvey's because it touched on what part of air was needed for life and that it seemed to be brought into the body by way of the lungs. This was part of a growing interest in "vitalism," the search for the spark of life that transformed inanimate matter into living plants and animals.

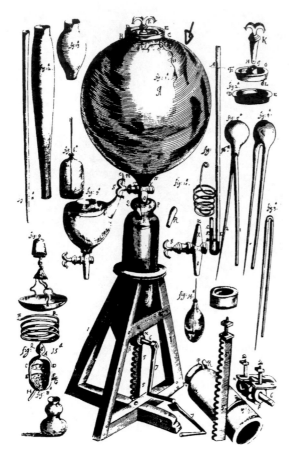

5.6 BOYLE'S AIR PUMP AND TOOLS FROM *NEW EXPERIMENTS PHYSICO-MECHANICALL* (1660)

Boyle's work also demonstrated the relationship between pressure (the "spring" of the title) and volume of air. He and Hooke used a j-shaped tube filled with mercury to show that increasing or decreasing the pressure on the short stem raised or lowered the level of mercury in the long stem. (See figure 5.6.) Boyle argued that "according to the *Hypothesis*, that supposes the pressures and expansions to be in reciprocal proportion";[1] in other words, as pressure goes up, the volume of air goes down in equal proportion and vice versa. While Boyle was making a specific argument about atmospheric air, which he considered an elastic fluid, not an element in itself, the basic relationship he pointed out was later transformed into $PV = K$, what we now call "Boyle's Law" or occasionally "Mariotte's Law" after Edmé Mariotte (1620–84) who independently found the same relationship in 1676.

Unfortunately, Boyle's air pump was plagued with problems. It leaked quite badly, so that it was not really possible to evacuate all the air from the interior. Also, although he published detailed accounts of his instrument and its operation, complete with schematic diagrams, other natural philosophers across Europe could not replicate his results. Thomas Hobbes (1588–1679), a natural philosopher as well as a moralist, severely criticized the work of Boyle and Hooke. Hobbes claimed that the air pump did not work and that it in no way represented a vacuum, as they had claimed. Although Hobbes presented many sound arguments, Boyle's rising status, both socially and scientifically, ensured that his was the winning side of this disagreement.

....

1. Robert Boyle, *A Defence of the Doctrine Touching the Spring and Weight of the Air* (London: M. Flescher, 1682) 58.

Despite equipment problems, the claim of replicability became fundamental to the experimental program from Boyle on. That is, natural philosophers began to claim that experiments were useful and produced true and certain results precisely because they were not dependent on the experimenter but could be repeated by anyone. The air pump was to be seen as transparent; nothing stood between the observer and the aspect of nature being observed. In other words, an ideology of the objectivity of instrumental experiment arose from Boyle's work with the air pump. Our modern reliance on experimentation—and on replicability—is in part based on the triumph of Boyle and the Royal Society over Hobbes and other skeptics. That "proof" (that is, evidence of the true state of nature) results from experimentation was, and is, a powerful idea, even if it has philosophical flaws. Many things make an unmediated view of nature impossible, but because of Boyle's status and the status of his witnesses, the air pump became "black-boxed." That is, the air pump and other experimental instruments were considered neutral and objective, unproblematically revealing nature's state. For example, when we look at a thermometer to help us decide what kind of coat to wear outside, we have accepted a particular concept of temperature. The thermometer is not an unbiased object but embodies a philosophical idea about the quantification of nature. To someone unfamiliar with the concept of temperature, a thermometer would be a meaningless device. This does not mean that what the thermometer reveals is false but that all scientific instruments represent a system of beliefs about the world.

The Development of Philosophical Instruments

During the seventeenth century, other instruments (often called philosophical instruments) and experimental programs were devised, all sharing this ideological stand with Boyle's work. Boyle himself used the barometer, developed in 1644 by Evangelista Torricelli (1608–47), and others followed this lead in investigating the weight of air. Torricelli filled different glass tubes with quicksilver (mercury), inverted them into a basin, and discovered that all the tubes maintained a constant level. Torricelli claimed that the space above the mercury, the "Torricellian space" as Boyle called it, was a vacuum and argued that "we live submerged at the bottom of an ocean of the element air, which by unquestioned experiments is known to have weight."[2]

....

2. Evangelista Torricelli, "Letter to Michelangelo Ricci" (1644), in *Encyclopedia of the Scientific Revolution*, ed. Wilbur Applebaum (New York: Garland, 2000) 647.

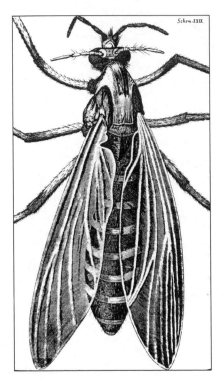

5.7 ILLUSTRATION FROM ROBERT HOOKE'S *MICROGRAPHIA* (1665)

He also predicted that if one ascended to higher altitudes, the weight of air would be less and the column of mercury would descend further. This prediction was taken up by the mathematician Blaise Pascal. Pascal first worked to replicate Torricelli's instrumental experiment, which proved a difficult task. He eventually succeeded, both with a column of mercury and with a much larger one of water. This resulted in a heated debate about the possibility of a vacuum, strongly denied by a number of leading theologians. To avoid this discussion and concentrate instead on the question of the weight of air, in 1648 Pascal took the barometer up a mountain near his brother-in-law's home in Clermont, France. Sure enough, the higher the ascent, the lower the column of mercury and the larger the "Torricellian space" at the top of the tube. Because Pascal soon thereafter had a crisis of faith and turned from natural philosophy to spirituality, the results of this investigation were not known until after his death, with the posthumously published *Traités de l'équilibre des liqueurs et de la pesanteur de la masse de l'air* (1663). This work, with Boyle's, set up an experimental research program for the coming century, as well as demonstrating the power and "objectivity" of philosophical instruments.

Perhaps the most innovative philosophical instrument of the seventeenth century was the microscope, first developed in the first decade of the century. Following the success of the telescope to bring distant sights closer, unknown instrument makers, probably in the Netherlands, produced instruments designed to greatly magnify the very small. The five microscopists best known for their startling observations and discoveries were van Leeuwenhoek, Hooke, Malpighi, Swammerdam, and Grew. Antoni van Leeuwenhoek (1632–1723), a merchant in Delft, first turned these magnifying devices on a variety of substances, most famously male semen, where he claimed to observe small animalcules in motion. This resulted in a series of interesting letters to the Royal Society. Robert Hooke turned to the favourite subjects of microscopists—insects, seeds, and plants—and captured some stunning images of various enlarged phenomena through engravings in his bestselling, lavishly illustrated book, *Micrographia* (1665). (See figure 5.7.)

The Italian anatomist Malpighi turned the microscope on the human body and, in addition to his embryological work, discovered capillaries and their role in the circulation of the blood. Jan Swammerdam of Amsterdam (1637–80) disproved contemporary theories about the metamorphosis of insects, while the Englishman Nehemiah Grew (1641–1712) found the cellular structure of plants. All five successfully overcame early suspicions that the instrument was creating and disguising as much as it was revealing to produce theories and observations much sought after by the natural philosophical community. By the eighteenth century the vanishingly small was seen to be observable, without scientists worrying about any interference from the apparatus itself.

Newton and the Experimental Method

Isaac Newton also played a part in the development of experiment (and experimental instruments) as a legitimate methodology. Beginning during his *annus mirabilis* he developed a theory of light based on a series of simple and elegant experiments. Because of a decades-long dispute with Hooke about optics, he refused to publish, but in 1703 Hooke died. Newton's *Opticks* came out in 1704. Unlike the *Principia*, the *Opticks* was written in English, in simple language, and laid out so that the experiments could be recreated by anyone who could read the book and afford a few pieces of optical equipment such as prisms, mirrors, and lenses. Even more than Harvey or Boyle, who had likewise explained their procedures in print, Newton became the model for the new experimental method. It was a smash hit, snapped up by an eager public and translated into French, German, and Italian within the year. It has been in print almost continuously to the present day.

Newton was contributing to a long tradition of optics research, stretching back to the Middle Ages and earlier Arabic natural philosophers. He was also building on the work of Kepler, who had argued that light travelled in rectilinear rays, enabling a mathematical description of its path. This allowed several natural philosophers, including Thomas Harriot (c. 1560–1621), René Descartes, and Willebrord Snell (1580–1626), to develop the sine law of refraction, which stated that when a ray of light passes from one transparent medium to another (such as air to water), the sine of the angle of the incident (original) ray divided by the sine of the angle of the refracted ray equals a constant. While this was a useful and demonstrable relationship, it led to a disagreement about the nature of light. Was it a motion through matter (that is, a wave)? Or was light made of particles? How did light travel? Could it travel in a vacuum? During the 1670s Huygens developed a wave theory of light in which he argued that the idea of a wave front represented the path of the light.

Newton criticized this wave theory, largely because it seemed to contradict the rectilinear nature of rays put forward by Kepler. Newton argued that light was corpuscular and proved it to his own satisfaction with a series of elegant experiments. By passing a beam of sunlight through a series of prisms, he demonstrated that white light was not pure light, as had been supposed, but rather was a composite of many colours (the spectrum). His demonstration has been called the *experimentum crusis* or crucial experiment, the demonstration that confirms the hypothesis. What Newton noticed was that light passing through a prism smeared into an oblong with colours at top and bottom. It had been assumed since antiquity that such an effect was the result of some corruption of the white light. If this was the case, it seemed reasonable to assume that by passing some coloured light through a second prism, the degree of corruption would be increased. So Newton passed light through a prism, then allowed a small amount of coloured light to project through a slit, and then through a second prism. There was no change in the colour of the light. In other words, nothing was added to or taken away (see figure 5.8.). To confirm this, he also placed two prisms together so that the first separated the light and the second, an inverted prism, gathered all the bands back together, producing a spot of white light.

Newton believed that light was composed of particles and that their differing speeds resulted in a differing angle of refraction when passed through a prism. If, for example, all red light was composed of small particles of similar nature, there seemed to be no mechanism to change the nature of the red particles as they passed through successive prisms.

This was hotly debated, the French following Descartes's and Huygen's lead in preferring a wave theory of light, but Newton's *Opticks* provided a foundation for a new English school of optical research throughout the eighteenth century.

Newton's *Opticks* also laid out a research program for natural philosophers who followed him. The book concluded with a series of "Queries," topics that interested Newton but which he had not had time to fully investigate. In addition, since there was a strong bias among Newton's peers against "theoretical" science, meaning the presentation of philosophical ideas without experimental demonstration, Newton presented his ideas as a series of questions. This list extended beyond optics, covering a range of scientific areas related to light such as the relationship between light and heat, the effect of the media of transmission on the behaviour of light, and the condition of the universe. For close to 100 years many natural philosophers and scientists interested in finding important areas of research started their inquiries

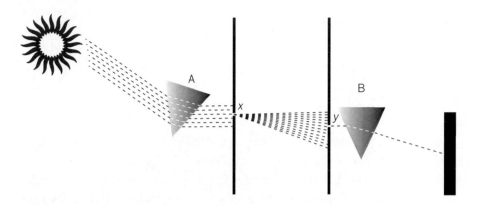

5.8 NEWTON'S DOUBLE PRISM EXPERIMENT

Sunlight passes through the prism "A" and falls on the screen. A portion of the spectrum produced (light of a single colour) passes through the screen at slit "x" and continues on to the second screen, where it goes through slit "y" and on to the second prism "B." The light is finally projected on to the wall. Newton's experiment demonstrated that light of a single colour could not be further broken up into a spectrum.

by taking on one of these questions. For example, in Query 18 (of the fourth edition of the *Opticks*), Newton noted that two thermometers, one in a vacuum and the other in a closed container of air, both seemed to heat up and cool down at about the same rate. This suggested to Newton that there must be a medium of propagation "more rare and subtle than the Air" that transmitted heat like a vibration. This observation prompted a number of scientists to look for the imponderable fluid or ether, an idea that was eventually resolved by the work of Einstein. It also got other scientists to investigate the nature of heat as separate from temperature, leading ultimately to the kinetic theory of heat and the laws of thermodynamics.

One of the most famous questions posed by Newton was Query 31 from the 1718 edition. Newton asked: "Have not the small Particles of Bodies certain Powers, Virtues, or Forces by which they act at a distance, not only upon the Rays of Light for reflecting, refracting, and inflecting them, but also upon one another for producing a great Part of the Phenomena of Nature?" Newton goes on to suggest that known forces such as gravity and electricity may play a role in the attraction of particles, but unknown forces may also be at work. By observing attraction, according to Newton, it should be possible to figure out the law of attraction that controls the way matter combines and functions. This idea contributed to the concept of chemical affinity, an idea used by most chemists to explain how matter combined. Affinity theory was one of the foundational concepts of modern chemistry until the nineteenth century when it was replaced by valence theory.

New Scientific Organizations

The new instruments, experiments, and the underlying assumptions about the relationship between humans and nature were fundamentally important to the creation of a new scientific enterprise in the seventeenth century. Equally important were the new institutional structures that developed in this period for the express purpose of fostering and supporting natural philosophy. The founding of such assemblies as the Royal Society of London and the Académie Royale des Sciences[3] in Paris contributed to a dramatic new organization of science, one that encouraged natural philosophers to develop social codes of behaviour, rules about who could do science and what counted as science, and statements about the secular usefulness of their enterprise. Just as Galileo had tried to separate religious and scientific claims to the monopoly of truth, so too did these new scientific societies. This was the beginning of the institutionalization of science, different in kind from the university-based science and distinct from the court-based science of the sixteenth century, although owing much to that earlier model. The breakdown of individual princely courts cast some natural philosophers adrift. Growing absolutism, especially in France, led to a more particular focus for patronage, as well as to a growing urban elite and intellectual culture in the capital cities. One hundred and fifty years of bloody religious wars caused people to look elsewhere than the church for salvational knowledge and for a secure group with status. The rising leisured class was looking for secular legitimation and something to do.

In about 1603 the first secular scientific society was formed, the Accademia dei Lincei (Academy of the Lynx) in Rome. This organization was founded by Federico Cesi (1585–1630), later Prince Cesi, and existed in great secrecy for the first years of its life, persecuted by the authorities including Cesi's father. Cesi was a man of huge intellectual energy and curiosity, who established a research program in natural history of great scope and imagination. When Galileo joined in 1611, the Accademia was given a major boost, especially since Galileo donated his microscope to it. After the condemnation of Copernicus, however, followed by Cesi's death in 1630 and Galileo's condemnation, the Accademia ceased to function. In 1657 a scientific society was founded in Florence, the Accademia del Cimento (of experiments). This society had neither a formal membership nor statutes and existed for ten years as a

---

3. The name of this organization changes according to informal or official use. Historians say "Académie des Sciences," although officially it is called "Académie Royale des Sciences." To make things more complicated, the name changed several times even though it remained the same group. Within the text, we will use the term commonly used by historians, that is, the Académie des Sciences.

loose collection of men interested in experimental research. It, like the Lincei, was in reality a hybrid, neither court-based nor autonomous, since it was focused on the court of Grand Duke Ferdinand II de Medici and his brother, Prince Leopold.

The most famous of the seventeenth-century scientific societies, the one that first marked this new form of scientific organization and the only one that has survived as a continuously operating body to the present day, was the Royal Society of London, established by royal charter in 1662. There is much debate about its origins. A number of people interested in natural philosophy, experimentalism, and the utility of natural knowledge met informally in England during the English Civil War and in the Interregnum that followed. The Civil War was fought between Royalists, who wished to maintain the power of the Crown and the liberal Anglicanism of the Church of England, and Parliamentarians, who argued for the primacy of political power from the people in Parliament and for the predestinarian theology of the Puritans. Historians of the Royal Society have sought founders of modern science in both camps, but most particularly among the Puritans. Samuel Hartlib (c. 1600–62), an educational reformer who came to England from Prussia to escape the Thirty Years' War and who was closely associated with the Parliamentary camp, tried to develop an educational program to bring natural philosophy and the new experimental method to English intelligentsia. The Hartlib circle was one of the groups responsible for the founding of the Royal Society after the Restoration, although on much more conservative and elite lines than Hartlib and his circle had envisaged. In fact, while earlier historians sought the origins of this new institutional structure in radical religion and politics, the truth seems to be that most natural philosophers were eager to find an alternative to the crippling political and religious controversies of their day and that natural philosophy provided just such a third way.

With the Restoration of Charles II in 1660, various groups from London, Oxford, and Cambridge came together in London to form the Royal Society. Although it received a charter from the king, it was autonomous and, therefore, different from earlier university, church, or court-based spaces of scientific investigation and discussion. It was founded with a strong inclination toward a Baconian philosophy of research, setting out to employ an inductive, cooperative method in order to discover useful information for the benefit of the nation. Here the rhetoric of utility that emerged with the courtly philosophers was carried into an urban, gentlemanly locale. Thomas Sprat, the first official historian of the Royal Society, said that the Society was founded as a way to avoid the "enthusiasm" of the Puritans and the sectarian disputes that had ripped the country apart during the Civil War. Although

Sprat was hardly a disinterested observer, it does seem that the Royal Society attempted to find a third way through the religious and civil disagreements of the period. While Royal Society members included those of many different religious affiliations (from Catholic to Puritan), what they had in common was the desire to keep clear of religious controversy and to do natural philosophy instead of theology.

The Royal Society developed a strict method of choosing members, who had to be known to existing members and have an active interest in natural philosophy. The exception to this were some aristocrats, necessary to maintain the Society's elite nature. It also refused to allow women to join, although Margaret Cavendish, Duchess of Newcastle (1623–73), did attend some meetings and had published more books on natural philosophy than many of the members put together. They were also very hesitant to accept tradespeople, preferring instead the trustworthiness of gentry. The Royal Society developed a gate-keeping function, determining who counted in natural philosophical

5.9 FRONTISPIECE OF SPRAT'S *HISTORY OF THE ROYAL SOCIETY* (1667)

inquiry. Through the work of their first corresponding secretary, Henry Oldenburg (c. 1619–77) and through the publication of their journal, the *Philosophical Transactions of the Royal Society*, founded in 1665 and still published today, they also were able to determine what counted as proper natural philosophical work. Thus, in one fell swoop, the Royal Society became the arbiter of just who could be a natural philosopher and what qualified as acceptable natural philosophy.

The other successful seventeenth-century scientific society was established in a very different way. The Académie Royale des Sciences was founded in Paris in 1666 by Louis XIV's chief minister, Jean-Baptiste Colbert. Although there had been an informal network of correspondence centred on Father Marin Mersenne (1588–1648) since the 1630s, the Académie was a top-down organization, another element of the absolutist French state. Unlike the Royal Society, where members were elected and unpaid, the state appointed 16 academicians, paid as civil servants, to investigate the natural world as the king and his advisors required. So the Académie can be seen as the root of the professionalization of science, since this was the first instance where scholars were paid exclusively as scientists. Because their research agenda was set by the state, they could take on projects beyond the scope of individual scientists. For example, the Académie sponsored the measurement of one minute of arc of the Earth's surface, resulting in the first accurate measurement of the size of the Earth and the distance to the stars. In the long run, however, it was less successful than the Royal Society. An appointment as an Academician was often the reward for a life's work rather than an incentive to new work. Most of the massive projects came to nothing. Its journal, *Journal des Sçavans* (founded in 1665), was largely a reprint service. The Académie des Sciences did well as a promoter of science as an elite and respected activity, but it was not a place that sponsored innovation.

These new scientific societies created four enduring legacies for science as a profession. First, science was now seen as a public endeavour, although with carefully defined limits, members, and methods. Second, its cooperative nature was stressed, through projects such as the History of the Trades sponsored by the Royal Society and the History of Plants and Animals investigated by the Académie. Such undertakings led to the Enlightenment view that all was knowable if properly organized and to the sense of the utility of the knowledge gained. When Leibniz founded the Berlin Academy of Science in 1700, he chose as its motto *theoria cum praxi*, theory with practice.

Third, scientific communications were established as an essential element of the scientific enterprise. While it had been true that letters between natural philosophers (for example, between Galileo and Kepler) or within the letter-writing circle of Father Mersenne had been integral to maintaining a community of scholars, the establishment of scientific journals in the seventeenth century both broadened and controlled this community. These journals acted as a guarantor of veracity and reliability, even while issues were socially determined and highly contested. They also broadcast scientific ideas and experiments to a much larger audience, allowing ordinary people to take part in science by "virtual witnessing." This led to a wider interest in the investigation of nature and a greater acceptance of new scientific ideas and of scientists as respectable, if awe-inspiring people.

Finally, scientific societies established scientists as experts, qualified by membership to pose and judge questions about nature. This was especially true in

France, where election to the Académie was the culmination of one's life's work. But equally within the Royal Society, some natural philosophers, such as Newton or Boyle, were respected both within and outside the society as experts and significant scholars. Below them were the collectors, those who found interesting natural phenomena to report but who left theorizing to their betters, much as Bacon had laid out in Solomon's House. Thus, the seventeenth-century scientific societies established ideologies about science and its practice still with us today.

During the sixteenth century, connections between natural philosophers and skilled artisans aided greatly in the development of new ideas about nature and new problems to be investigated. This was also true during the seventeenth century, although the focus became the urban mercantile centre, rather than the princely court. Just as the scientists themselves were forming associations with their fellow scholars in these new, secular, non-court settings, often in major trading centres, they were also closer to shipyards, print shops, instrument makers, and chart makers. However, projects like the Royal Society's History of the Trades, an unmitigated disaster, shows us that the communication between these artisans and scholars was more complicated than one might think. The History of the Trades attempted to find out how all the different manufacturing trades in England were performed, so that natural philosophers might find a more rational scientific way to manufacture goods. Not surprisingly, the tradesmen were remarkably unforthcoming about their trade secrets, and the suggestions made by the bewigged gentlemen were at best unhelpful and at worst positively dangerous. It would take some time for a new collaborative approach to bring together the skills of craftspeople and the precision of scholars.

The Place of Women in the Study of Science

Most of this discussion of seventeenth-century science has focused on the contributions of men of science; this was, however, a period when women were attempting to make their mark on the study of the natural world. Both Anne Finch, Viscountess of Conway (1631–79), and Margaret Cavendish, Duchess of Newcastle, were interested in breaking into this previously clerical and male preserve. The same social and intellectual upheaval that made science a gentlemanly pursuit gave women a brief window of opportunity to become involved in natural philosophy. Bethsua Makin (c. 1612–c. 1674) wrote *An Essay to Revive the Ancient Education of Gentlewomen* (1673), arguing for the right and ability of women to study natural philosophy. Anne Conway corresponded with Leibniz and shared with him her theory of "monads" that became the basis of his particulate philosophy of the universe. Noblewomen such as Christina of Sweden engaged in natural philosophical conversations. Margaret Newcastle wrote many books of natural philosophy and attended a meeting of the Royal Society. She also composed what has been called the first English work of science fiction, *The Description of a New World, Called the Blazing World* (1666). These women, however, were exceptions. The seventeenth and eighteenth centuries saw restrictions on women's sphere in science as in much else.

Such restriction came from a general change in attitude to a gendered nature and from changing theories about women's role in society and in reproduction. In a pre-industrial society the vast majority of people existed in a close and symbiotic relationship with nature, which was perceived as female, a nurturing mother. The ideal was coexistence, not control. Mining, for example, was either the rape of the Earth or the delivery of a child, since minerals developed in the Earth's womb. Therefore, sacrifices, prayers, and apologies were necessary before mining could begin. The vitalism of Paracelsus, the natural magic of the neo-Platonists, and the naturalism of Aristotle all gave the female contribution of nature and to nature its due.

In early modern Europe, however, this began to change. As scholars began to view the Earth as exploitable, its image changed to a wild female who must be tamed. At the same time women were losing socio-economic status, becoming less autonomous and less able to earn wages or operate independently in craft guilds. Increasingly, accusations of witchcraft were levelled, especially against women. Scientific theories of sexuality and procreation changed. Where during the Middle Ages women had been recognized as providing something to initiate procreation (matter for Aristotle, female semen for medieval authors), during the sixteenth and seventeenth centuries theorists such as Harvey claimed that women were merely a receptacle, the incubator of offspring, and, as such, were totally passive. Likewise, women were no longer seen as equal partners in intercourse but rather seducers, enticing men into sexual activity that was deleterious to their health and intellectual well-being. Women were also losing their professional role in reproduction, as female midwives were replaced by licensed male surgeons or male midwives, who used their new technology of forceps to manage nature. Finally, with the introduction of the Mechanical Philosophy, introduced partly to deal with perceived disorder in the world, the soul of nature was stripped away, leaving only inanimate atoms; nature was dead, and women's claim to vitality through reproduction was rendered void. In other words, the natural philosophy of the seventeenth century articulated an ideology of exploitation, an image of the world that could be constructed according to man's specifications.

We can trace this change in attitude toward women as natural philosophers through the careers of two scientists: Maria Sybilla Merian (1647–1717) and Maria Winkelmann (1670–1720). Their careers demonstrate, on the one hand, the possibility of women's involvement in natural inquiry and, on the other, the restrictions to their participation through the creation of the new institutions of scientific societies.

Maria Merian's career illustrates the success a woman could have in a scientific field, particularly one based on an entrepreneurial model. Merian was born in Germany into a family of artists and engravers. From an early age she was interested in drawing and painting insects and plants. After marrying her stepfather's apprentice she became a renowned insect and plant illustrator, publishing well-received and beautifully engraved books. In 1699 the city of Amsterdam sponsored her travels to Suriname, where she observed and recorded many new plants and animals. On her return she created a bestselling book of these findings, which was published posthumously. Merian's career thus followed a very successful path in the older apprenticeship and mercantile model. However, after her death her work came to the attention of the new community of natural historians and philosophers, and her reputation suffered. Her Suriname book was strongly critiqued, condemned for her classification system and more particularly for her credence of slave knowledge about the use of these plants and insects. Her reputation, therefore, was greatly diminished in the increasingly misogynistic and racist attitudes of the eighteenth-century scientific community.

Maria Winkelmann was the daughter of and later wife of astronomers (she married Gottfried Kirch in 1692) and worked with both her father and husband on telescopic observations in Berlin. In 1702 Winkelmann independently discovered a comet and published her findings. She was a full participant in her chosen scientific field, but after Kirch's death in 1710 her status fell sharply. The Royal Academy of Sciences in Berlin refused to allow her to continue in her husband's position as official astronomer to the Academy, eventually appointing her less capable son instead. Even Leibniz's support was insufficient to help her maintain her position; the Academy was unwilling to set a precedent by allowing a woman to hold such an important job. In this they followed the lead of the Royal Society, which had likewise debated allowing Margaret Cavendish to join and had resisted the undesirable precedent. Eventually the Academy, embarrassed by her presence at the observatory, forced her to leave the premises; without access to large telescopes, she was unable to continue her observational work. The new organization of science, the Academy, proved itself to be more restrictive for women than the earlier apprenticeship model had been.

Seventeenth-Century Scientific Ideology

The new ideology of exploitation of and superiority toward nature reflected a changing attitude toward knowledge and nature. As crucial as the new knowledge and methodologies proved to be, of even greater significance in the creation of modern science were the new locales of scientific discussion and the new ideology and code of conduct the seventeenth-century societies established. Science had previously been the property of clerics and academics, but the upheaval of the seventeenth century—its religious and political wars, its economic strife—created an opportunity for a new group of gentlemen practitioners to develop a new standard for scientific conduct and a new place to practise science. This was particularly true in England, where the move away from absolutism at the end of the century allowed a certain freedom of association among the leisured classes and where the controversies associated with the Civil War and its aftermath gave gentlemen a desire for civility and an alternate way to establish matters of fact.

Robert Boyle was particularly instrumental in this transformation of scientific ideology. First, he set up laboratories, spaces for scientific experiment and investigation, in private locations, particularly in his sister's townhouse in fashionable Pall Mall in London. The privacy of this space was paramount, since Boyle was able to control access and behaviour within the site. That is, he could allow in people with the proper credentials, worthy to witness his experiments and guaranteed to know the proper way to behave. Similarly, private museums of natural history developed in this period all over Europe as private spaces, also belonging to aristocrats and gentry, who controlled visitors and bestowed on those visitors status through their permission to observe.

Because this space was private, Boyle could decide who had the proper credentials to observe, to take part in experiments, and to participate in the making of natural knowledge. He developed a number of criteria, later used by the Royal Society and other scientific bodies. The person seeking entry had to be known to Boyle or his circle and, thus, was a gentleman. This person should also be a knowledgeable observer, one who was able to validate the experimental knowledge and to witness matters of fact, rather than just gawk. Still, it was more important that the observer be of the leisured classes than that he or she be philosophically knowledgeable. This was clear from von Guericke's famous hemisphere experiment, which gained its status as one that created natural knowledge about the weight of the air by the presence of a large group of gentry onlookers (as can be seen in figure 5.5). Similarly, one of the important roles of the

SIR ISAAC NEWTON.

5.10 NEWTON

Engraving of Newton. From Sarah K. Bolton. *Famous Men of Science*. NY: Thomas Y. Crowell & Co., 1889.

Royal Society was to provide a credible audience for various experimental demonstrations, thereby establishing their veracity.

Because these experiments and demonstrations of knowledge were performed in spaces created and used by gentlemen, gentlemanly codes of behaviour were adopted as the codes of behaviour for scientists. For example, gentlemen argued for the openness and accessibility of private space at the same time that they were carefully controlling access to such places. Likewise, scientific laboratories as they developed claimed to be open public space while limiting access to those with the knowledge and credentials to be there. Gentlemen were very concerned with issues of honour and argued that their word was their bond. A gentleman never lied, which was why there was much outrage (and duels fought) at any suggestion of cheating. This was why matters of fact could be established through the witnessing of a few gentlemen, who would, of course, see the truth of the investigation and report it accurately to others. A scientist's word was also his bond. Scientists would not cheat or lie, and thus they claimed the role of a completely trustworthy enclave in society. The Royal Society in concert with Boyle created a community of scientists who could decide what constituted truth and reality and who were allowed to make pronouncements on that reality.

Conclusion

By the time Newton died in 1727, the place of the natural philosopher had changed considerably. The image of the natural philosopher had shifted even further away from the "pure" intellectuals of ancient Greece, the Islamic wise men, or even the courtiers of Galileo and Kepler's era. A bust commemorating Newton sits in the entrance hall of Trinity College, Cambridge. (See figure 5.10.) It does not present him as an erudite scholar or as an important member of society, as had earlier portraits from his student days, or as Master of the Mint. Rather, it presents him as a modern Caesar who, with firm gaze and noble brow, has conquered all he surveyed. While Caesar captured Rome and gained an empire, Newton conquered

Nature and made it man's dominion. The poet Alexander Pope said of Newton's life: "Nature and Nature's Laws lay hid in night / God said, 'Let Newton be,' and all was light."[4]

By the end of the seventeenth century many aspects of modern science had been established. Philosophers of this period of scientific revolution had wrestled with questions of epistemology at the beginning of the century and decided on a behavioural model of truth-telling by the end. Some of these thinkers developed new ideas, theories, and experimental discoveries, setting in place a series of research programs for the coming century. They had also introduced a new methodology of science, which included the mathematization of nature and a new confidence in, and reliance on, experimentation. New secular scientific institutions sprang up; with them came an articulation of the utility of their knowledge to the state and the economy. Finally, the domination of natural philosophical enquiry by secular gentlemen from the leisured classes ensured a code of behaviour that was gendered and class-based. These ingredients led to a new scientific culture that rapidly assumed a recognizably modern face. All these changes together constituted a scientific revolution.

Essay Questions

1. Was there a scientific revolution? If so, of what did it consist?

2. In what way can Newton be seen to have completed the Copernican revolution in astronomy?

3. In what ways were the Royal Society and the Académie des Sciences similar and different?

4. What role did women play in the development of natural philosophy and did this change over time? Why or why not?

---

4. Alexander Pope, "Epitaph Intended for Sir Isaac Newton." In Westminster Abbey, 1735.

FURTHER
READING

ONE   *THE ORIGINS OF NATURAL PHILOSOPHY*

Aristotle. *Meteorologica*. Trans. H.D.P. Lee. Cambridge, MA: Harvard University Press, 1952.

Aristotle. *Physics*. Trans. Hippocrates G. Apostle. Grinnell, IA: Peripatetic Press, 1980.

Aristotle. *Posterior Analytics*. Trans. Jonathan Barnes. Oxford: Clarendon Press, 1975.

Bernal, Martin. *Black Athena: The Afroasiatic Roots of Classical Civilization*. New Brunswick, NJ: Rutgers University Press, 1987.

Byrne, Patrick Hugh. *Analysis and Science in Aristotle*. Albany, NY: State University of New York Press, 1997.

Clagett, Marshall. *Greek Science in Antiquity*. Freeport, NY: Books for Libraries Press, 1971.

Irby-Massie, Georgia L., and Paul T. Keyser, eds. *Greek Science of the Hellenistic Era: A Sourcebook*. London: Routledge, 2002.

Lloyd, G.E.R. *Early Greek Science: Thales to Aristotle*. London: Chatto and Windus, 1970.

Lloyd, G.E.R. *Greek Science after Aristotle*. London: Chatto and Windus, 1973.

Lloyd, G.E.R. *Magic, Reason and Experience: Studies in the Origin and Development of Greek Science*. Cambridge: Cambridge University Press, 1979.

Lloyd, G.E.R., and Nathan Sivin. *The Way and the Word: Science and Medicine in Early China and Greece*. New Haven, CT: Yale University Press, 2002.

Plato. *The Republic*. Trans. G.M.A. Grube. Indianapolis, IN: Hackett Publishing, 1992.

Plato. *Timaeus*. Trans. Francis M. Cornford. Indianapolis, IN: Bobbs-Merrill, 1959.

Rihll, T.E. *Greek Science*. Oxford: Oxford University Press, 1999.

Tuplin, C.J., and T.E. Rihll, eds. *Science and Mathematics in Ancient Greek Culture*. Oxford: Oxford University Press, 2002.

Zhmud, Leonid. *Pythagoras and the Early Pythagoreans*. Trans. Kevin Windle and Rosh Ireland. Oxford: Oxford University Press, 2012.

TWO   *THE ROMAN ERA AND THE RISE OF ISLAM*

Baker, Osman. *The History and Philosophy of Islamic Science*. Cambridge: Islamic Texts Society, 1999.

Beagon, Mary. *Roman Nature: The Thought of Pliny the Elder*. Oxford: Oxford University Press, 1992.

Bricker, Harvey M., and Victoria R. Bricker. *Astronomy in the Maya Codices*. Philadelphia: American Philosophical Society, 2011.

Dallal, Ahmad S. *Islam, Science, and the Challenge of History*. New Haven, CT: Yale University Press, 2010.

French, Roger, and Frank Greenaway, eds. *Science in the Early Roman Empire: Pliny the Elder, His Sources and His Influence*. London: Croom Helm, 1986.

Glasner, Ruth. *Averroes' Physics: A Turning Point in Medieval Natural Philosophy*. Oxford: Oxford University Press, 2009.

Harley, J.B., and David Woodward. *History of Cartography, Volume II, Book I. Cartography in the Traditional Islamic and South Asian Societies*. Chicago: University of Chicago Press, 1992.

Hogendijk, J.P. *The Enterprise of Science in Islam: New Perspectives*. Cambridge, MA; London: MIT Press, 2003.

Huff, Toby E. *The Rise of Early Modern Science: Islam, China, and the West*. Cambridge: Cambridge University Press, 1993.

Lehoux, Daryn. *What Did the Romans Know? An Inquiry into Science and Worldmaking*. Chicago: University of Chicago Press, 2012.

Masood, Ehsan. *Science & Islam: A History*. London: Icon, 2009.

Principe, Lawrence M. *The Secrets of Alchemy*. Chicago: University of Chicago Press, 2013.

Qadir, C.A. *Philosophy and Science in the Islamic World*. London: Routledge, 1990.

THREE   *THE REVIVAL OF NATURAL PHILOSOPHY IN WESTERN EUROPE*

Brotton, Jerry. *The Renaissance Bazaar: From the Silk Road to Michelangelo*. Oxford: Oxford University Press, 2002.

Grant, Edward. *The Foundation of Modern Science in the Middle Ages, Their Religious, Institutional and Intellectual Contexts*. Cambridge: Cambridge University Press, 1996.

Grant, Edward. *Planets, Stars and Orbs: The Medieval Cosmos 1200–1687*. Cambridge: Cambridge University Press, 1994.

Grant, Edward, ed. *A Source Book in Medieval Science*. Cambridge, MA: Harvard University Press, 1974.

Kibre, Pearl. *Studies in Medieval Science: Alchemy, Astrology, Mathematics and Medicine*. London: Hambledon, 1984.

Lindberg, David C. *The Beginnings of Western Science: The European Scientific Tradition in Philosophical, Religious, and Institutional Context, 600 B.C. to A.D. 1450*. Chicago: University of Chicago Press, 1992.

FOUR   *SCIENCE IN THE RENAISSANCE: THE COURTLY PHILOSOPHERS*

Biagioli, Mario. *Galileo, Courtier: The Practice of Science in the Culture of Absolutism*. Chicago: University of Chicago Press, 1994.

Biagioli, Mario. *Galileo's Instruments of Credit: Telescopes, Images, Secrecy*. Chicago: University of Chicago Press, 2006.

Blair, Ann. *The Theater of Nature: Jean Bodin and Renaissance Science*. Princeton, NJ: Princeton University Press, 1997.

Bono, James J. *The Word of God and the Languages of Man: Interpreting Nature in Early Modern Science and Medicine*. Madison: University of Wisconsin Press, 1995.

Cormack, Lesley B. *Charting an Empire: Geography at the English Universities, 1580–1620*. Chicago: University of Chicago Press, 1997.

Daston, Lorraine. *Wonders and the Order of Nature, 1150–1750*. New York: Zone Books, 1998.

Drake, Stillman. *Galileo: Pioneer Scientist*. Toronto: University of Toronto Press, 1990.

Finocchiaro, Maurice A. *Defending Copernicus and Galileo: Critical Reasoning in the Two Affairs*. New York: Springer, 2010.

Galilei, Galileo. *Dialogue Concerning the Two Chief World Systems— Ptolemaic and Copernican*. Trans. Stillman Drake. Foreword by Albert Einstein. Berkeley: University of California Press, 1967.

Galilei, Galileo. *Two New Sciences, Including Centers of Gravity and Force of Percussion*. Trans. Stillman Drake. Madison: University of Wisconsin Press, 1974.

Gingerich, Owen. *The Book Nobody Read: Chasing the Revolutions of Nicolaus Copernicus*. New York: Penguin Books, 2004.

Grafton, Anthony, with April Shelfor and Nancy Siraisi. *New World, Ancient Texts: The Power of Tradition and the Shock of Discovery*. Cambridge, MA: Belknap Press of Harvard University Press, 1992.

Magnus, Albertus. *The Book of Secrets of Albertus Magnus of the Virtues of Herbs, Stones and Certain Beasts, also A Book of the Marvels of the World*. Ed. Michael R. Best and Frank H. Brightman. Oxford: Clarendon Press, 1973.

Moran, Bruce T., ed. *Patronage and Institutions: Science, Technology, and Medicine at the European Court, 1500–1750*. Rochester, NY: Boydell Press, 1991.

Newman, William Royall, and Anthony Grafton, eds. *Secrets of Nature: Astrology and Alchemy in Early Modern Europe*. Cambridge, MA: MIT Press, 2001.

Saliba, George. *Islamic Science and the Making of the European Renaissance*. Cambridge, MA: MIT Press, 2007.

Swerdlow, Noel, and Otto Neugebauer. *Mathematical Astronomy in Copernicus's* De Revolutionibus Part 1–2. Studies in the History of Mathematics and Physical Sciences 10. New York: Springer-Verlag, 1984.

Westman, Robert. *The Copernican Question: Prognostication, Skepticism, and the Celestial Order*. Berkeley: University of California Press, 2011.

Vollmann, William T. *Uncentering the Earth: Copernicus and* On the Revolutions of the Heavenly Spheres. New York: Norton, 2006.

FIVE   *THE SCIENTIFIC REVOLUTION: CONTESTED TERRITORY*

Bala, Arun, ed. *Asia, Europe and the Emergence of Modern Science: Knowledge Crossing Boundaries*. New York: Palgrave, 2012.

Dear, Peter Robert. *Revolutionizing the Sciences: European Knowledge and Its Ambitions, 1500–1700*. Princeton, NJ: Princeton University Press, 2001.

Harkness, Deborah. *The Jewel House: Elizabethan London and the Scientific Revolution*. New Haven, CT: Yale University Press, 2007.

Hunter, Lynette, and Sarah Hutton, eds. *Women, Science and Medicine 1500–1700: Mothers and Sisters of the Royal Society*. Stroud, UK: Sutton, 1997.

Jardine, Lisa. *Ingenious Pursuits: Building the Scientific Revolution*. New York: Nan A. Talese, 1999.

Lindberg, David C., and Robert S. Westman, eds. *Reappraisals of the Scientific Revolution*. Cambridge: Cambridge University Press, 1990.

Long, Pamela O. *Artisans/Practitioners and the Rise of the New Sciences, 1400–1600*. Corvallis: Oregon State University Press, 2011.

Newman, William Royall. *Atoms and Alchemy: Chymistry and the Experimental Origins of the Scientific Revolution*. Chicago: University of Chicago Press, 2006.

Newton, Isaac. *Opticks*. New York: Prometheus, 2003.

Newton, Isaac. *The Principia*. Trans. Andrew Motte. New York: Prometheus, 1995.

Osler, Margaret J. *Reconfiguring the World: Nature, God, and Human Understanding from the Middle Ages to Early Modern Europe*. Baltimore, MD: Johns Hopkins University Press, 2010.

Osler, Margaret J. *Rethinking the Scientific Revolution*. Cambridge: Cambridge University Press, 2000.

Park, Katherine, and Lorraine Daston, eds. *Early Modern Science. The Cambridge History of Science*, vol. 3. Cambridge: Cambridge University Press, 2003.

Schiebinger, Londa L. *The Mind Has No Sex?: Women in the Origins of Modern Science*. Cambridge, MA: Harvard University Press, 1989.

Shapin, Steven. *The Scientific Revolution*. Chicago: University of Chicago Press, 1996.

Smith, Pamela H. *The Body of the Artisan: Art and Experience in the Scientific Revolution*. Chicago: University of Chicago Press, 2004.

Westfall, Richard S. *Never at Rest: A Biography of Isaac Newton*. Cambridge: Cambridge University Press, 1980.

GENERAL READINGS IN THE HISTORY OF SCIENCE

Asimov, Isaac. *The History of Physics*. New York: Walker, 1984.

Bowler, Peter J., and Iwan Rhys Morus. *Making Modern Science: A Historical Survey*. Chicago: University of Chicago Press, 2005.

Brock, William H. *The Norton History of Chemistry*. New York: Norton, 1992.

Bronowski, Jacob. *The Ascent of Man*. Boston: Little, Brown, 1973.

Bryson, Bill. *A Short History of Nearly Everything*. Toronto: Anchor Canada, 2004.

Diamond, Jared. *Guns, Germs, and Steel: The Fates of Human Societies*. New York: W.W. Norton, 2005.

Golinski, Jan. *Making Natural Knowledge: Constructivism and the History of Science*. Cambridge: Cambridge University Press, 1998.

North, John David. *The Norton History of Astronomy and Cosmology*. New York: Norton, 1995.

Olby, Robert C. *Fontana History of Biology*. New York: Fontana, 2002.

INDEX

Illustrations indicated by page numbers in italics